Understanding

HI-FI CIRCUITS

BY NORMAN CROWHURST

ORIGINALLY PUBLISHED BY
GERNSBACK LIBRARY INC.
1957

KCK Media Corp.
Publisher

First Reprint Edition
2002

ISBN: 1-882580-38-9

Printed in the United States of America

contents

introduction

THE whole subject of high fidelity has "come of age" in one or two short decades. With such rapid progress along such varied avenues, some confusion was to be expected. Also the assessing of hi-fi performance is so much a matter of taste, while its circuitry is generally regarded as completely scientific. Whatever may be under discussion, each enthusiast naturally claims that his favorite, whether it be a whole system, an amplifier, preamplifier, or even a part of a circuit, is the best.

The hi-fi enthusiast is by nature a perfectionist. But getting the best is not so simple a matter as putting together all the "best" parts. A high fidelity system must be "integrated," or considered as a whole. To questions often asked, there is no such simple answer as "The split load type is the best phase inverter." But neither is the story terribly complicated. This book divides the whole audio or high fidelity system into sections, and explains the "facts of life" pertinent to each section. *Your* selection for the best of a particular circuit should depend on your choice of output stage and the feedback performance you expect. This, in turn, depends upon the kind of amplifier you want. This principle is equally true, whether you buy the amplifier complete, or make it yourself.

In most sections there are at least two approaches—one simple and one complex—each with some merits over the other. The best compromise will depend on a number of circumstances: your budget, your taste in programs, how critical a listener you are, and in what way. This book aims at helping you unlock the answers to all these often-perplexing questions for yourself, whether you are an engineer, a technician, a musician, or just someone who would like to enjoy good music of your own choice.

Happy listening.

NORMAN H. CROWHURST

chapter 1

the approach to hi-fi

SERIOUS audio design dates back about three decades. At that time the design of audio circuits was in the stage of breaking away from being just part of a radio receiver or transmitter. Specialists were beginning to make audio amplifiers for phonographs and public-address systems. Fidelity was, of course, nothing to be compared with modern possibilities and the early methods of "development" would be frowned upon by present engineering standards.

In those days there was little more to choose from in the way of tubes than a few battery-operated triode types of varying impedance, or mu—according to which term of reference you may prefer. Most of the people "in the business" were virtually hobbyists and the kit radio, to construct at home, was very popular.

The radio sets of about that time contained a great many triode tubes; a number of tuned circuits at radio frequency, followed by an "audio amplifier" consisting of a number of tubes and interstage transformers to get enough amplification to give an audible level from the comparatively feeble transmitters then operating.

Early advances

Then the screen-grid tube or tetrode first made its appearance and shortly afterward the pentode. These came as an innovation and were hailed as producing fabulous results. Now you could get "much more power" from the sets. "It gave real punch" and similar rather nonscientific terms were used.

Some, however, noticed that the triode, while it did not give quite so much output, gave *cleaner* reproduction. The pentode gave more "punch" but at the expense of quality. There was some distortion which the triode did not give. Some interstage transformers behaved much better than others in this regard too.

And thus the quest for improved quality in audio began. Nonlinear distortion is, largely, a function of how much headroom

you have. That is, if you want to listen to 5 watts of audio power, the peaks may run up to 25 or 50 watts. So an amplifier of this handling capacity must be used to avoid distortion. This explains the quest for larger output stages with good performance to achieve clean output.

Push-pull operation reduces second-harmonic distortion which, in those days, was the major variety. It also enables the output transformer to be designed much more easily and with improved performance. But this created a problem—a phase inverter is necessary, unless you use push-pull operation from the beginning to the end of the amplifier. Some designers of those days favored this arrangement.

In the early days of reproduction from phonograph records some kind of tone control may have been used, but professional equalization was not visualized. Rather, adjustments were made until the program material just sounded nice.

How confusion arose

A great variety of people, working at almost an equal variety of technical levels, have achieved vast progress since those days. But some of these people have been working in ways that have confused the score for the uninitiated.

For one example, this writer knew some English "engineers" whose principal hope of fame then rested in acquiring copies of American magazines and finding clever ideas to apply in audio circuits. If nothing appropriate could be found in magazine articles, these individuals would start scribbling on paper or try connecting components in different places in actual circuits, in an endeavor to do "something clever."

The desire to do *something clever* still persists, under the impression that a clever circuit will gain an edge on one's competitors.

While some followed this method, many workers in the field concentrated on getting something which works practically, without their knowing quite why. Yet others went overboard in the theoretical direction and hired mathematicians to take care of the theory. This writer remembers one instance where the design of an acoustic horn for a loudspeaker was being investigated.

The mathematician hired for the purpose spent days going through involved mathematical functions and eventually came up with the solution: that the particular horn design in question had a *negative* acoustic resistance. To the mathematician this statement apparently had little meaning. But, to the practical engineer who heard it, it seemed as if we had struck across an acoustic form of perpetual motion—because a negative resistance must in essence be a generator!

The only kinds of negative resistance realized in practice require a B-plus supply—and usually filament as well—to provide the necessary energy for the generator. A negative acoustic resistance, therefore, should oscillate, all by itself, without any input. This would be very convenient because, by appropriate matching, one could make a horn loudspeaker with an overall efficiency exceeding 100%!

Of course, somewhere during those days of figuring, the mathematician had made a mistake!

Those panaceas

But in audio literature the trend to look for the "ultimate," or perfection, or some special trick which will prove to be a panacea for all audio troubles, has persisted. As an example, the cathode follower was once hailed as being the answer to all audio problems. It provides a high input source resistance, multiplying the natural input characteristics of the tube by its operating gain. It provides a low output source resistance which is very helpful for many purposes. It also operates at negligible distortion.

However, if you regard the cathode follower as a matching device to provide a low output source for matching into a low load, it definitely doesn't do all of these things at the same time. It may provide the low output source very nicely, but connecting the low value load to it removes all the feedback that you *thought* you had and thereby destroys its effectiveness in the other spheres just mentioned.

You never get something for nothing, not even in audio circuitry. In this example, the cathode follower provides impedance transformation and distortion reduction. If you take advantage of the impedance transformation for matching purposes, the distortion reduction disappears. If you take advantage of the distortion reduction and a convenient low source impedance, then

you cannot utilize its matching property by connecting a low value load to it.

Reaching the ultimate

An aspect of this philosophy of trying to do something that achieves the "ultimate" in audio can be illustrated by the use of positive and negative feedback. The development of this method arose due to the large loss of gain that is necessary, with negative feedback, to achieve a great amount of improvement in performance.

To reduce distortion by a factor of 100 to 1 it is necessary to use 40-db feedback. This in turn means that 40 db of gain must be thrown away. Then it was discovered that the gain could be increased artificially, as well as being reduced, by the use of *positive* feedback, rather like regeneration in radio. So, by using positive feedback over a small section of the amplifier, the gain could be increased. Then negative feedback could be employed over the whole amplifier to achieve an improved reduction in distortion as illustrated in Fig. 101.

A "logical" ultimate to this would appear to be the condition where positive feedback is used to provide a gain of infinity. This

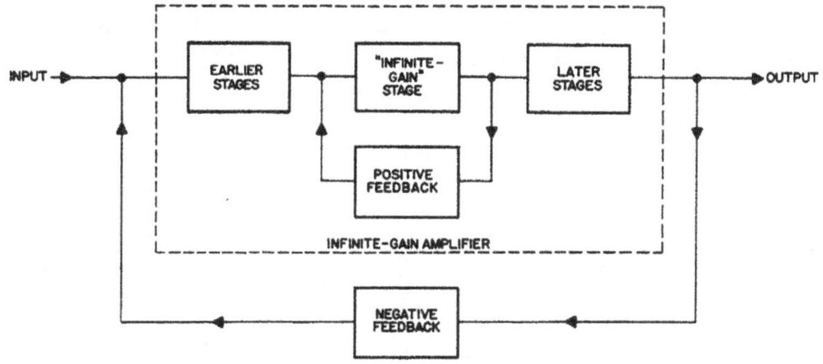

Fig. 101. *Two-loop feedback amplifier, showing attainment of "infinite feedback."*

occurs when the signal fed back over the small section of amplifier is exactly equal to the original input. This means that the overall amplifier to which negative feedback is applied will also have infinite gain.

So it would seem that we should come out with zero distortion and a perfectly flat frequency response because now *the use of any feedback at all* becomes *infinite* feedback.

Have we at last found a way of getting "something for nothing"? It might appear so, until we realize that the question of which feedback loop is the outer one depends upon viewpoint. We can redraw the circuit of Fig. 101 as at Fig. 102, showing the same block elements.

Here it appears that the rest of the amplifier and the negative feedback loop can be regarded as an alternative feedback path over the single stage, in parallel with the positive feedback. This means that application of negative feedback destroys the infinite gain property of this stage.

In reality the stage can be rendered stable again after positive feedback has made it oscillate, merely by connecting the overall negative feedback. In the chapter on feedback more examples of this kind of interaction will be considered and the subject will be discussed in full to show how feedback can affect performance.

It is introduced at this point merely as an illustration of the kind of misleading presentation that has been made from time to time. The implication the reader draws is of an *ultimate* achievement that will revolutionize audio.

Confusion compounded

The purpose of this book is, not to go into the details of design in the various stages of audio circuits, but rather to answer the question "What's the score?" So many methods have been introduced, in almost every section of audio, that the uninitiated finds himself confused. This confusion has, in fact, very often affected the engineer's approach to his problem through no fault of his.

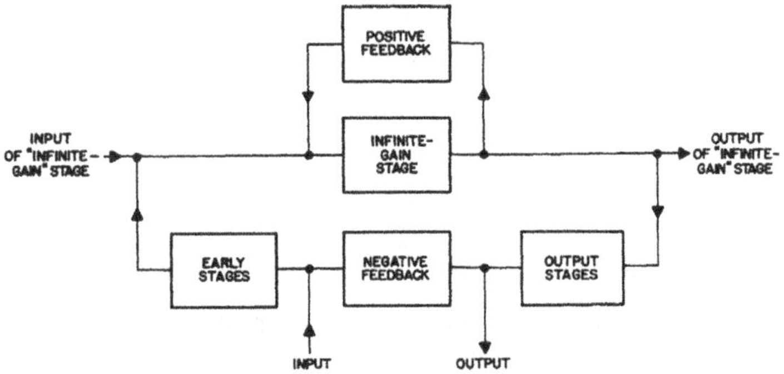

Fig. 102. *Block diagram of Fig. 101 in rearranged form.*

To illustrate, a difference is sometimes noted between the performance of two circuits or amplifiers which is the opposite of that anticipated. A circuit which, according to measurements and design, should give better performance than another, is observed to bear the opposite relationship—the inferior circuit sounds better. Maybe it's an ultra-linear output circuit that, according to measurements and theory, should give a superior performance to either a triode or pentode-type output circuit. However, why should a particular tryout seem to show the opposite?

This improved performance is dependent upon the ultra-linear transformer behaving correctly at all frequencies throughout the audio band. A practical transformer may behave perfectly *at all frequencies used for test purposes* and yet introduce spurious effects on program material. This is not a direct reflection on the ultra-linear circuit; it is something that did not occur in either the triode or pentode circuit. So it is possible that a particular ultra-linear circuit may sound inferior to either of the older circuits, although, according to theory, it is supposed to be superior.

This does not invalidate the theory; it merely shows that all the facts have not been taken into account. But the uninitiated inevitably would judge the ultra-linear circuit adversely. This happens time and again in audio.

Often this kind of misjudgment has given the engineer additional problems because he has to satisfy the people who evaluate his work (whether this is the ultimate public or just the people who employ him) and these people are apt to believe their ears and completely disregard the engineer's theorizing!

There is an opposite philosophy the designer has had to contend with too: the continual strife for better specifications. This approach reverses the procedure and puts the theoretical specifications ahead of practical performance.

The endeavor for improved performance initially showed that the flatter the frequency response the better an amplifier would sound. Furthermore, the wider the frequency response extended, at least up to the full width of the audio band, the better the amplifier sounded.

Also an amplifier with only 5% distortion sounds considerably better than one with 10%.

This being the case, the natural endeavor was to improve the bandwidth and flatness of the response and to reduce the distortion as indicated by measuring equipment.

In the specifications of the amplifier, however, this process has been taken to extremes. Measurements have been made from time to time to determine just how much deviation in frequency response, and also how much distortion, the ear can detect. Most modern high-fidelity amplifiers, according to measurement and specification, give distortion that is only a small fraction of the smallest that can be detected audibly.

According to this criterion then, all high-fidelity amplifiers should sound identical. In practice however they don't—even when used with the same phono, pickups, tuners and loudspeakers. The sound given by different amplifiers varies, although the difference in specification is so small as to prove the difference inaudible. What does it mean?

Things happen in amplifiers that are not described in the simple technical specifications or found by the simple method of measurement. Amplifiers have ways of misbehaving that do not show up under conventional measurement methods and it is this misbehaving that makes the difference audible.

In this book, as well as discussing the features of the varieties of circuits available for each purpose that contribute different results—in terms of standard specifications and performance—we shall also endeavor to give details and descriptions of the things that can happen in the circuit that do not get described by these technical specifications. In this way we will give a more complete picture of the relative merits of circuits so that the reader may decide for himself at each stage which is the best circuit for his particular purpose.

The chapter arrangement may seem unusual. The philosophy behind its arrangement is intended to follow the theme set here. The first big problem in getting good sound is a satisfactory output stage. So we start there and work our way through the progressive obstacles to the approach to an "ultimate" in high fidelity.

In reading this book, nowhere will you find a direct statement that such and such is the best kind of amplifier—or even the best circuit for this part of an amplifier. In the complications of audio circuitry, no such unequivocal statement is possible. Every circuit will have its advantages and its disadvantages. The determination of which is the best circuit for an individual purpose will depend entirely on the relative importance of the various advantages and disadvantages for the particular job in hand.

To aid in seeing, at a glance, the relative characteristics of the circuits considered, several of the chapters conclude with a tabulation of the merits and demerits of the varieties considered in that chapter.

The book does not stop at circuits, because these are not all there is to audio. We listen to sound waves, not electrical impulses. So a complete consideration of the best in audio must consider every link from the original sound to the listener's ears.

chapter 2

special output stages

A DECADE or so before this book was written, the score in this part of audio circuits would have been a simple one to discuss. However it was nonetheless strongly debated. The choice was then a simple one, between triode or pentode operation of the output tubes.

Some contenders were in favor of what might be termed brute-force operation, using "enough triode" to get the required available output. The feature stressed about triode output was that, although it was less efficient than the pentode output stage, it was inherently lower in distortion.

The contenders for the pentode output stage pointed to the higher efficiency of this kind of tube and showed that the same number of watts could be achieved at considerably lower cost. They also maintained that pentode distortion is no higher, and sometimes lower, than that of triodes.

Both sides had their points, which is largely why the variety of circuits we shall discuss in this chapter came into being—in an endeavor to achieve the advantages of both. Today so many different output stages are available and the claims put forward for them are so difficult to reconcile one with another that many, engineers even, are confused as to their relative merits and the essential features of each.

To get a clear picture, the best method is to take a particular pair of output tubes, such as the 5881 (about which we have plenty of data) and compare their operation in the various circuits possible. Other tube types may vary slightly the relative merits of circuits using them but, in general, different tubes will just alter the magnitude of output.

As far as available power and basic distortion are concerned we need only to consider three methods of operation: triode, tetrode (or pentode) and ultra-linear. The other variations in this chapter can easily be derived from one or more of these modes of operation by using partial or full cathode-follower action or some other rearrangement of the circuit in which the tube is used.

There is another variant that needs consideration: the kind of power supplies used, particularly for the bias. The way the amplifier behaves on transient loud passages, compared with sustained passages, often depends on the design of the tube supplies. These incorporate large capacitors. For a short burst of power, the capacitors will not have time to discharge appreciably so the power delivered will have the same supply voltage to draw from as at low power. But for a sustained passage, the increased plate current will cause the plate supply voltage to drop gradually. In a well-designed amplifier, two things are necessary for uniformly good handling of program material:

a) The changes in supply voltages that occur due to level increase must not appreciably affect either the distortion or gain of the amplifier. Should they do so, the initial part of a loud passage will sound different from its "follow through."

b) The changes must be arranged to occur at a uniform rate so that no peculiar effects arise in transition. *All the supply voltages (screen, plate and bias) must have the same time constant.*

Whatever circuit is used, the plate current of an output stage increases with signal level. This means the supply voltage will drop, unless special arrangements are made for good regulation—which are usually not economic. At the same time the bias voltage increases: in the case of automatic bias, due to increased plate current through the bias resistor; in the case of fixed bias, due to grid current when clipping occurs.

Careful design will allow a circuit using automatic bias to fulfill both the conditions described, but fixed bias presents problems that often are not satisfactorily solved for program presentation. When clipping occurs, the negative bias is increased due to grid current pulses. The bias supply has a relatively long time constant to smooth out ripple in a "no-current" circuit. So when clipping ceases, the increased negative bias stays longer than the change in plate voltage.

Fixed bias will generally get a larger maximum output from a given pair of tubes by allowing more favorable operating conditions to be used. This real increase in output may not always be apparent. To compare maximum output, volume is usually turned up to see "how much it will give." So clipping will occur in each amplifier. With fixed bias, erratic changes of bias occur that cause distortion not present when the level is kept within the limit. With automatic bias, the change is only evident for the duration of the clipping, which is audibly more acceptable. Thus it happens that the listener will often conclude that the automatic bias amplifier gives a bigger output than the one with fixed bias, although measurements show the reverse.

Push-pull triodes

Fig. 201 shows the circuit for push-pull triode operation with fixed bias. This requires a separate bias supply but has the advantage that a greater output is available from the pair of tubes. Fig. 202 illustrates the composite tube characteristics for a pair of 5881's operated in this manner, using a plate supply of 400 volts and −45 volts of fixed bias. Shown dashed are the individual characteristics of each tube, while the solid lines joining the dashed curves indicate the composite characteristics of the tubes operating in push-pull.

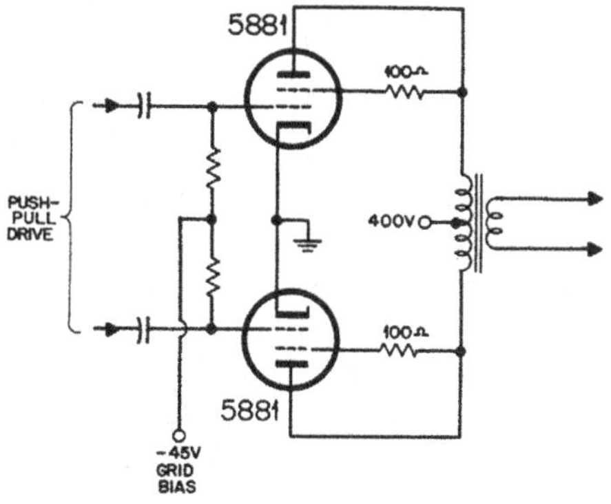

Fig. 201. *Circuit for pair of 5881 tubes operating in push-pull as triodes with fixed bias.*

The central solid load line represents the optimum load condition—a plate-to-plate load of 4,000 ohms. This operating condition yields an output of 13.3 watts with 4.4% distortion which analyzes into about 4% third harmonic and 1.5% fifth.

The 4,000-ohm plate-to-plate load line is somewhat of an ideal. It is the condition under which amplifiers are usually measured but unfortunately they do not normally work under this ideal.

The load resistance may vary, either above or below 4,000 ohms, and, what is worse, it often contains additionally some reactance. Fig. 203 shows two sets of curves for triode operation.

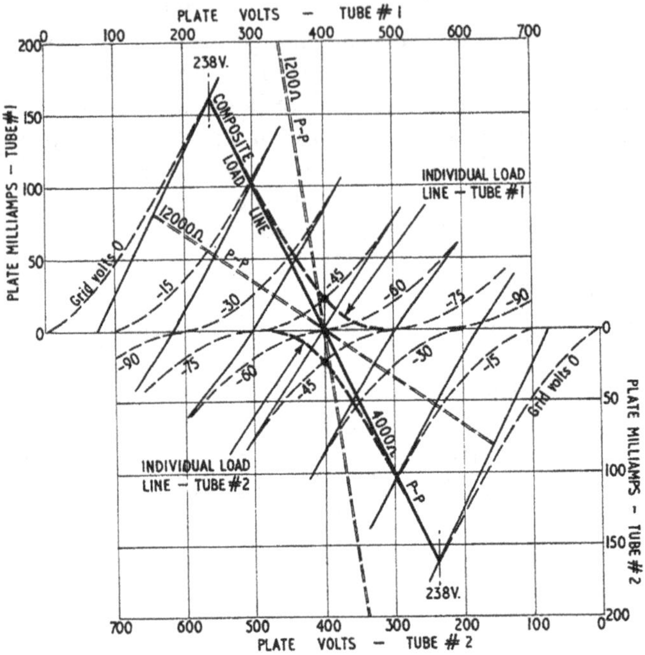

Fig. 202. *Composite curves for the working conditions shown in Figure 201. The significance of the various dashed lines is explained in the text.*

One pair shows the variation in distortion at maximum power and also the variation in the maximum power available before clipping occurs. A resistance load value is used in all cases, but its value is varied between 0.3 and 3 times the optimum value of 4,000 ohms plate to plate. These limits are represented by the dashed lines in Fig. 202.

But practical loads for amplifiers, which are usually speakers, with or without crossover networks, also incorporate reactances which cause the load line to become elliptical as represented in Fig. 204. The other pair of curves in Fig. 203 shows the variation in distortion at maximum power and the maximum power before clipping occurs under a hypothetical load condition made up as follows: a basic resistance load of 4,000 ohms plate-to-plate to which reactance is added. In one direction a shunt reactance decreases the load *impedance* as far as 0.3 of the 4,000 ohm plate-to-plate optimum and in the other direction a series reactance increases the total load impedance up to 3 times its optimum value.

When these curves are compared with the corresponding ones (given in Fig. 208) for pentode operation, it will be easy to see why the triode type output was preferred.

These curves have all been taken for push-pull operation with fixed bias. Sometimes, for economy in circuit design, automatic bias is used with the circuit of Fig. 205. Fixed bias is preferable because the combined plate current fluctuates with the signal. With 45-volts bias and no signal the plate current is 65 milliamperes, but when the output is driven to the full length of the load line, the current rises to 130 milliamperes.

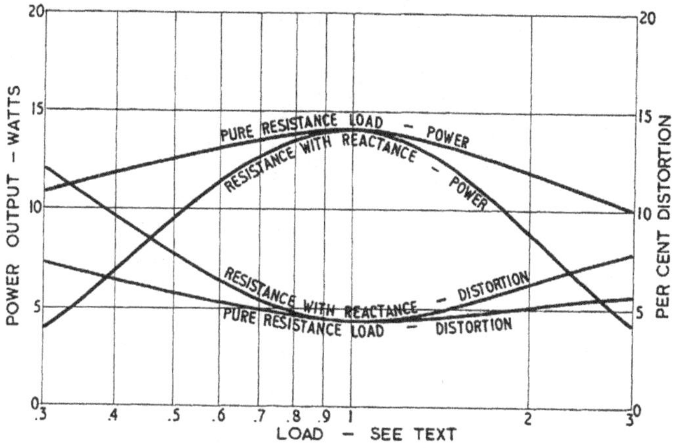

Fig. 203. *Output characteristics for push-pull triode operation as loading is varied.*

Suppose the bias had been obtained by the automatic bias circuit of Fig. 205. Then at the zero-signal condition a current of 65 ma would need to be fed through a resistor of about 700 ohms to produce about 45 volts; at maximum-signal condition the current of 130 ma would need to be fed through a resistor of only 350 ohms to produce a drop of approximately 45 volts. If a resistor of 700 ohms is used, then under maximum-signal condition the bias would rise to 90 volts. This would result in distortion due to the tubes being driven back beyond cutoff. On the other hand, if a resistor of 350 ohms were used, the bias at zero-signal condition would only be 22.5 volts. This would result in overheating the tubes when there is no signal or during quiet passages.

A further complication arises from the fact that in an automatic bias circuit, the bias voltage is subtracted from the total B-plus supply. This means that, if the bias swings from 30 to 60 volts and the B-plus supply is 440 volts, the plate-to-cathode voltage will swing from 410 to 380.

A self-bias circuit must choose a different operating condition for the tube, such that the plate current does not change quite

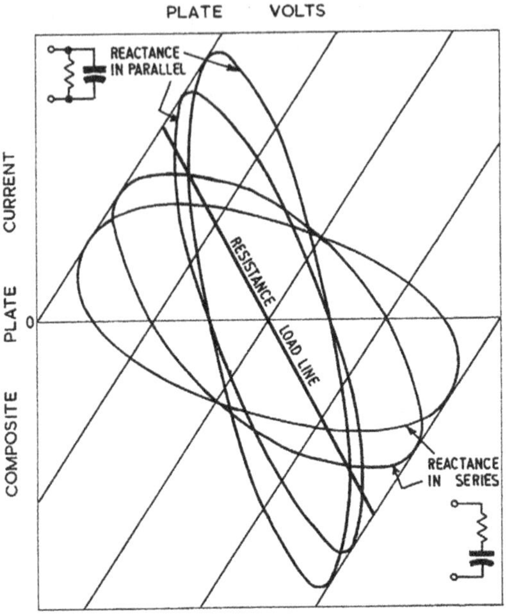

Fig. 204. *Construction of a load line representing reactance combined with the original resistance.*

so much from zero to maximum signal. A typical condition for self-bias operation of 5881's as triodes is: 400 volts plate-to-cathode; 35 volts bias (supply = 435 volts); plate current 130 ma; bias resistor 270 ohms; plate-to-plate load 8,000 ohms; power output 8.2 watts at 5% distortion, mostly low order third.

No mention has been made of "class" of operation: class A, class

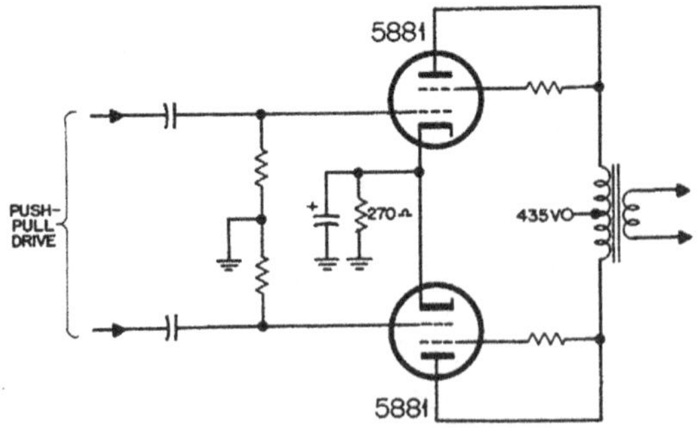

Fig. 205. *Circuit for pair of 5881 tubes operating in push-pull as triodes with automatic bias.*

B, class AB, etc. When these terms were first used, they provided a ready means of distinguishing between different operating conditions. Soon after they were introduced it became necessary to add subscripts to indicate how far the tubes were driven in the opposite direction. Thus class AB_1 would indicate that the tubes are so biased that for small signals the operation is class A—essentially linear on both tubes—while for maximum signal each tube is cut off for part of the cycle and that the tubes are not driven into the positive grid region to get maximum output.

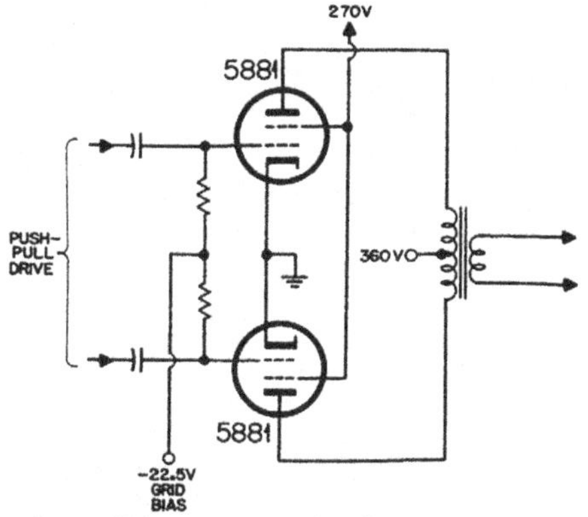

Fig. 206. *Circuit for pair of 5881 tubes operating as pentodes with fixed bias.*

What was not appreciated until later was the fact that changing the operating condition permitted changes in the fundamental concept of load matching: a triode could be matched into a load approximately equal to its plate resistance instead of several times higher. But a closer examination of the load line shows that this change, made to tubes set up for class-A operation, results in the same bias point yielding class AB operation. So now, to describe completely the class of operation, we need to specify what the loading does as well as the relative operating bias. The "simplified" designation has become as complicated as the details it is intended to replace. So we believe the best information can now be given in terms of operating voltages, currents and loading values.

Tetrode or pentode

Turning now to tetrode or pentode operation, the circuit is shown in Fig. 206 and the composite load condition in Fig. 207, using a screen voltage of 270 with a plate voltage of 360. These are the conditions quoted in the tube manual. A plate voltage of 400 could be used, which would emphasize still further the difference in available power between triode and pentode operation. Using the quoted plate-to-plate resistance of 6,600 ohms, the power output is 26.5 watts or just double that when the tubes are operated as triodes. With 400 volts on the plates it should be possible to get 35 watts from this method of operation.

The distortion quoted in the tube manual is 2%, which closely agrees with that measured in practice. However, there is an im-

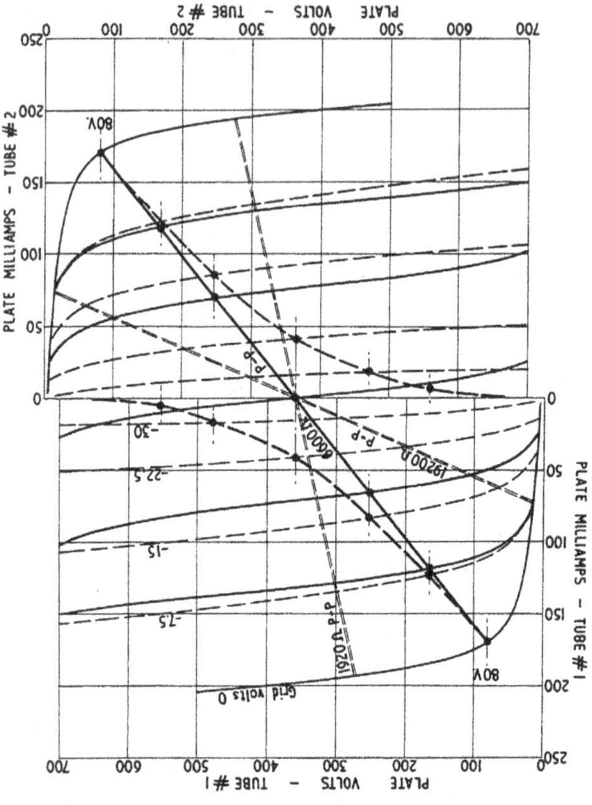

Fig. 207. *Composite curves for the working conditions shown in Fig. 206*

portant difference to be noticed between this distortion and that produced by the triodes. Analysis of the 2% shows it to be approximately 1.7% third harmonic and 1% seventh. The 1% seventh harmonic can contribute more annoying distortion than the 1.5% fifth harmonic present in the triode operation, because the spurious tones it causes—both harmonic and IM—are much more discordant.

When we turn to the distortion–power output characteristics of Fig. 208 however, we see something else about pentode operation. It is much more critical, not only of the load resistance used being closely in compliance with the optimum value, but distortion and maximum output are also critically dependent upon absence of reactance. With a reactance shunting the resistance down to an impedance of 0.3 times optimum, the pentode gives less output than the triode arrangement although, under optimum conditions, the pentode output is about double that of the triode arrangement.

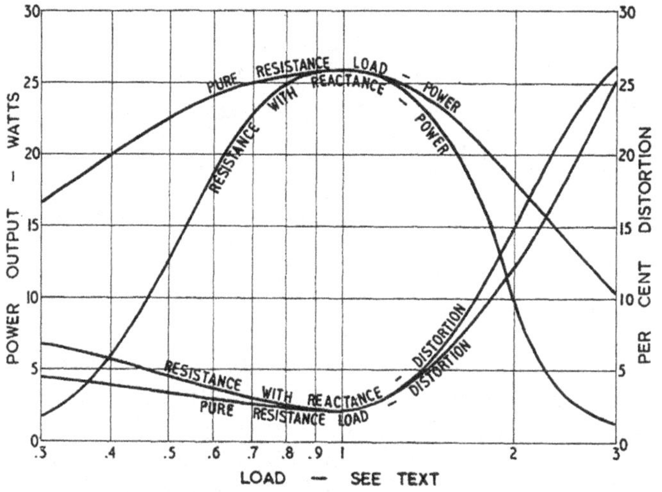

Fig. 208. *Output characteristics for push-pull pentode operation as loading is varied.*

A 3-to-1 change from optimum value is by no means unusual in speaker impedance characteristics. In fact the change is often greater than this. From this it is evident that although, according to the methods used for measuring output and distortion, the pentode gives about twice the output available from triode operation, yet in practical working conditions, the triode gives just as much useful power into a speaker load as the pentode.

Ultra-linear

This is why the ultra-linear operation came into being. Using as we have, a tetrode—the 5881—to make the equivalent of either a triode or pentode, we can see that:

To operate as a triode the screen, or No. 2 grid, swings with the same potential as the plate.

To operate as a pentode the No. 2 grid is kept at constant potential while the plate is allowed to swing.

Ultra-linear operation splits the difference: the screen grid swings, but by a voltage less than that of the plate. This is achieved by the use of a tapping on the primary of the output transformer, as shown in Fig. 209. Static tube characteristics for this method of operation, as published by Tung-Sol, are illustrated in Fig. 210. These curves are obtained by changing the screen voltage in a different manner from the plate voltage to simulate the way in which the plate and screen would swing away from the operating point of 400 volts.

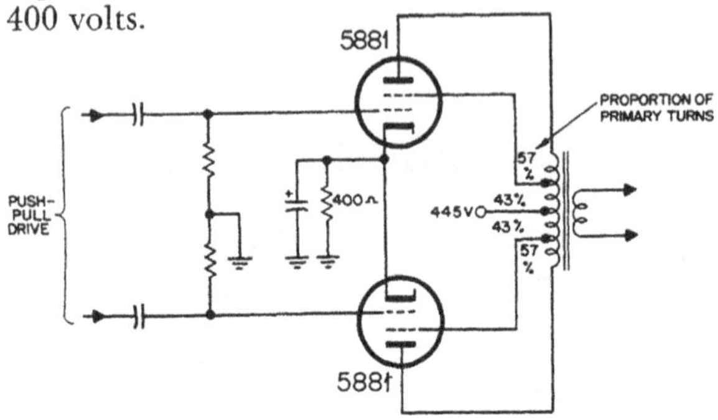

Fig. 209. *Circuit for pair of 5881 tubes operating in ultra-linear arrangement.*

With the ultra-linear circuit, the change of total plate current with amount of drive is not so great as with other methods of operation. With this particular tube, there is little advantage in fixed-bias operation for this reason. The operating condition for maximum drive is –45 volts. If this is obtained by automatic bias, it only drops to –40 volts in the quiescent condition. This observation does not apply to all tube types. For some it is advantageous to use fixed bias—a bigger output can be obtained within the plate dissipation rating of the tubes.

Using a plate-to-plate load of 6,500 ohms and a plate supply of 445 volts, the maximum power obtainable from a pair of 5881's in ultra-linear is 28 watts, with a distortion of about 3.3%, almost pure third. This is a very useful compromise between triode and pentode operation.

Pentode operation might give 35 watts with 400 volts on the plate—it would not be permissible to use 400 volts on the screen because the tube would be overrun. Notice the improved characteristics shown in Fig. 211. Although for wide deviations from optimum load ultra-linear operation does not hold up as well as the triode, for smaller deviations it is about as good as a triode, while it always gives better results than a pentode.

If only the optimum load condition were considered, it would appear that the triode gave the highest distortion, with 4.4%, while the pentode gave the lowest with 2%. But when deviation from optimum load, such as occurs in practice, is considered, it is found that the ultra-linear consistently gives the lowest distortion, with the triode running second and the pentode quite a poor third. It is only by the critical operation of the pentode that the low figure of 2% is obtained and then an analysis of the content of this 2% shows that it is of relatively high order.

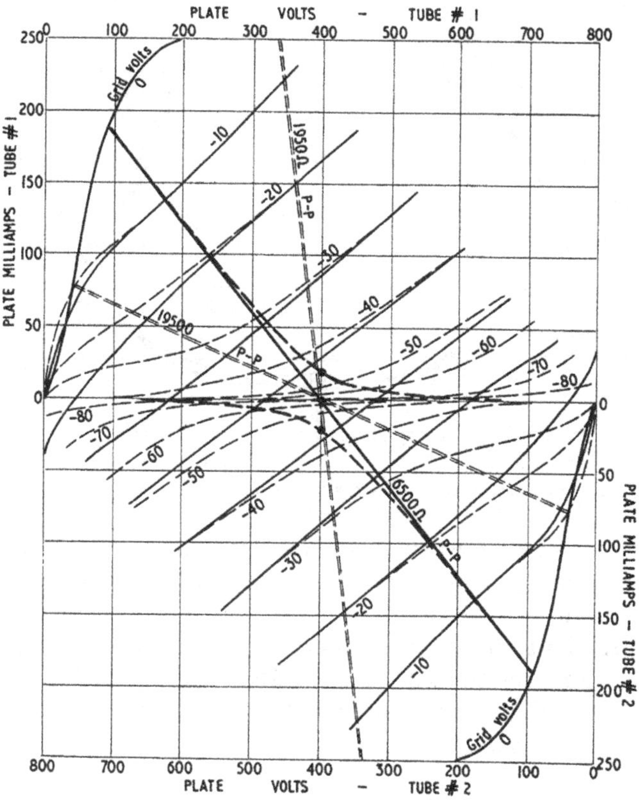

Fig. 210. *Composite characteristics derived from the curves published by Tung-Sol for 5881's in ultra-linear operation.*

Grid drive

So far we have discussed only the power output and distortion of these different circuits. In designing a whole amplifier we also have to consider the voltage drive necessary at the grids to give this output. In more complicated circuits the voltage drive needed is modified by the use of partial or wholly cathode coupling.

Grid drive is specified in a number of different terms. In referring to a tabulation of tube data, take care to ascertain which value it is that is listed. Some give the rms voltage per grid; some give the rms voltage from grid to grid (of a push-pull stage) which is twice this value; some give the peak voltage on each grid, which is 1.414 times the rms voltage; others give the peak-to-peak voltage, which is double this. This latter is really the most useful figure, because it gives the best picture of the drive required. Each grid requires this peak-to-peak voltage in opposite phase but the

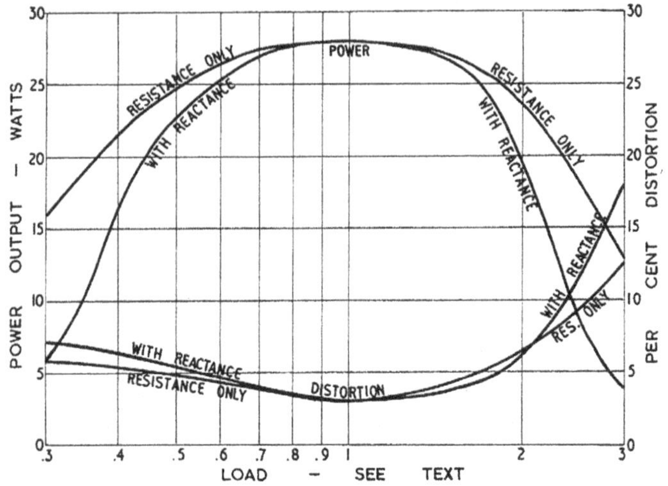

Fig. 211. *Output characteristics for ultra-linear operation as loading is varied.*

maximum voltage between grids (ac-wise) is the same peak-to-peak value because at the center of both swings they are at the same potential.

In the straight circuits just discussed the triode operation requires 90 volts peak-to-peak on each grid, the pentode circuit 45 volts peak-to-peak on each grid and the ultra-linear condition also 90 volts peak-to-peak on each grid.

Effective plate resistance

Closely associated with the matter of grid drive is the effective ac resistance the stage presents to the output terminals. Matters such as damping factor and feedback are discussed in the next chapter. We will see when we come to that point, that there is a limit to what can be done. This means we should also compare the starting point provided by the methods of operation so far discussed.

The straight triode operation of the 5881, with a 4,000-ohm plate-to-plate load resistance and 45 volts fixed bias, gives an average ac resistance in the plate circuit of 5,600 ohms or about 1.4 times the load resistance. This means that, on the secondary of the output transformer, a 10-ohm speaker would be feeding from an effective source resistance of 14 ohms, due to the plate resistance of the tubes.

Some readers may wonder why the triode resistance is as high as 1.4 times the load resistance. The usual value given is in the region of 0.3. This is based on the older method of operating triodes, in class-A, when the ac resistance is usually a fraction of the load resistance. For example, in the automatic circuit, using the values given for Fig. 205, the source resistance is about 3,200 ohms plate-to-plate, which is 0.4 of the load resistance. This means that a 10-ohm loudspeaker matched to this circuit would be fed from a source resistance of 4 ohms.

In pentode operation the effective ac resistance is between 5 and 10 times the load resistance but it is not constant throughout the output cycle. This is an additional reason why feedback is necessary for pentode stages: to linearize the ac resistance of the output stage.

The ultra-linear circuit, working with a plate-to-plate load of 6,500 ohms gives an ac source resistance of about 1.25 times the load resistance. This means a transformer matching a 10-ohm loudspeaker would provide a source resistance of 12.5 ohms, in the absence of feedback to modify this.

From this comparison the ultra-linear again shows up as the best starting point because it has the lowest ac resistance compared with the load resistance (except for the triode automatic bias circuit which gives only 8 watts output).

Cathode followers

The simplest variant from straight triode, pentode or ultra-linear operation is to use the tube as a cathode follower. This means that the plate is effectively grounded in place of the cathode and the output power is taken from the cathode circuit instead of the plate. Actually, of course, the tube still needs B plus to operate it so this means the plate is returned solidly to B plus instead of through the output transformer.

Fig. 212 shows the cathode-follower circuit for 5881's operating as triodes. Turning back to the tube characteristics of Fig. 202, the plate swing is from the operating voltage of 400 down to 238 for a 45-volt grid swing in each direction. Under the quiescent condition there is 400 volts from plate to cathode and the grid is 45 volts negative from the cathode. When the grid swings 45 volts positive from this—or to the same potential as the cathode—the cathode swings 162 volts positive from its normal potential, which means the grid drive must swing a total of 162 + 45 = 207 volts positive from its normal potential.

So the triode cathode-follower circuit will require a peak-to-

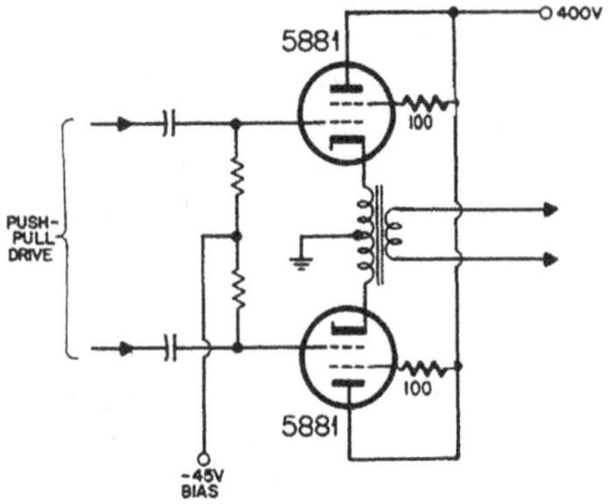

Fig. 212. *Circuit for operating push-pull triodes as cathode-follower output.*

peak swing of 414 volts for each tube. The power output will be the same as in straight triode push-pull, but let's see what happens to the distortion.

Of the 207 volts excursion applied to the grid, the 162-volt component from cathode to ground will contain 4.4% distortion of the 45-volt component from grid to cathode. But the total voltage of 207 is undistorted because it is the input voltage applied. So this means that the 45- and 162-volt components will each contain a distortion element in opposite phase.

If the 45-volt component possessed 4.4% harmonic, the 162-volt component would be pure, and vice versa. So the harmonic components will be about 3.4% of the 45 volts and 1% of the 162 volts, each of which amounts to 1.6 volts. But the 162 volts is the power output voltage. So this means that this method of operation has reduced the distortion from the region of 4.4% to about 1%. The source resistance provided by the cathode-follower triode operation is 1,250 ohms, referred to the cathode-to-cathode winding, for which the load was 4,000 ohms. This is a source resistance of 0.31 times the load resistance.

Pentodes may be operated in a cathode-follower arrangement by using the circuit of Fig. 213. This requires a transformer with an extra winding because the screen has to swing with potential variations identical to the cathode. In this case the input swing required is 280 volts for the cathode-to-ground output plus 22.5 volts for grid to cathode, a total of 302.5 volts.

Using the same reasoning as for the triode, this will reduce the distortion figure from 2% to about 0.15%. Under this condition there will be an effective source resistance of 450 ohms cathode-to-cathode which is .068 times the nominal cathode-to-cathode load of 6,600 ohms. Thus there is some advantage to the pentode cathode follower compared with the triode cathode follower: first

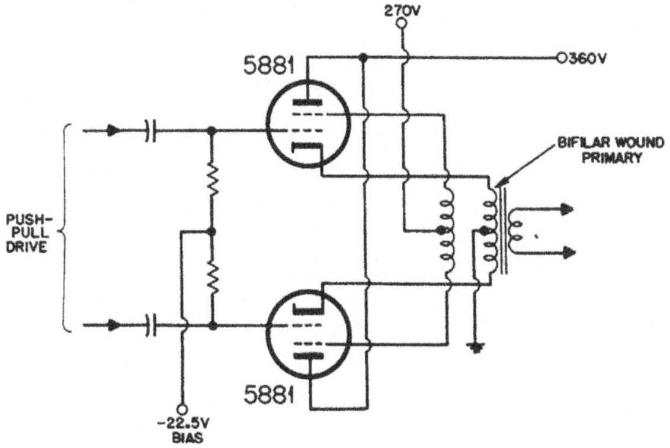

Fig. 213. *Circuit for operating as pentode cathode-follower output.*

with regard to distortion and second in reduction in source resistance. But this is at the expense of a greater swing—302.5 volts per grid as against 207. And remember, the pentode connection

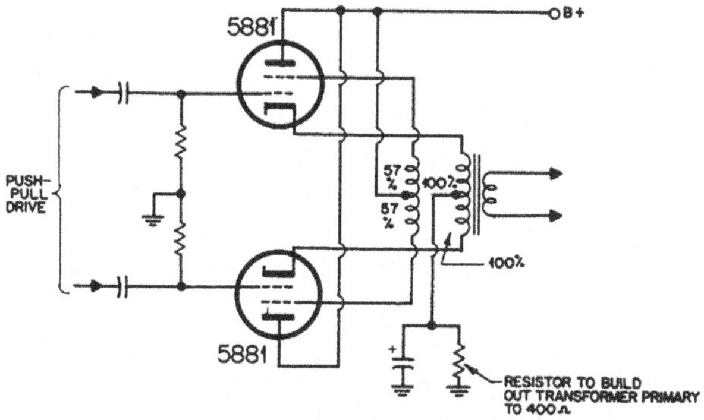

Fig. 214. *Circuit for operating as ultra-linear cathode-follower output.*

uses a plate voltage of only 360 compared to 400 for the triode.

It would also be possible to operate the ultra-linear circuit as a cathode follower by coupling the screen so that its voltage swings by 57% of the cathode swing while the plate voltage is connected to B plus, as shown in Fig. 214. This would increase the required grid drive to 345 volts peak or 690 volts peak-to-peak, which is even more drive than that required for pentode operation. The distortion will be reduced to less than 0.5% and the source resistance will appear to be about 1,100 ohms cathode-to-cathode or about one-sixth of the load resistance.

Full cathode-follower operation in either of the three major circuits is not generally used in practice because it is as difficult to obtain (without distortion) the very large drive voltages required as to design the output stage itself with low distortion. However there are a number of circuits which are virtually "halfway" toward the cathode follower so consideration of cathode coupling provides a good basis for understanding the other combinations.

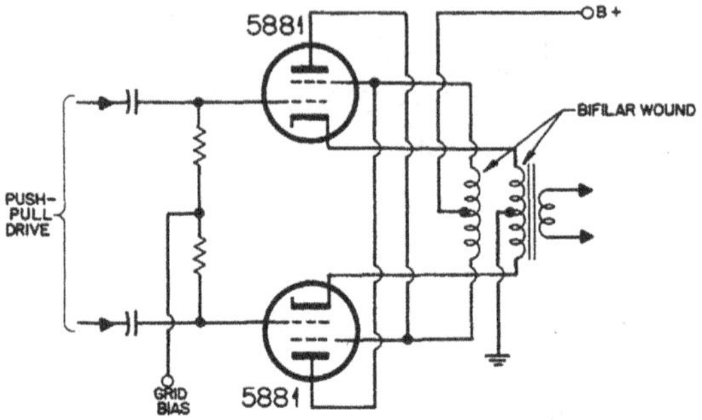

Fig. 215. *The unity-coupled circuit.*

Unity coupling

The first of these is unity coupling. The circuit for this is shown in Fig. 215. It is virtually a pentode circuit which is a halfway cathode follower. This means that the plates, screens and cathodes all swing by equal amounts but the phasing is such that each screen swings in exactly the same way as the corresponding cathode while the plate swings in the opposite direction.

For both the pentode cathode follower and this circuit, the design of the output transformer is extremely important. It is vital that the screen shall be tightly coupled to its corresponding cathode, otherwise instability will occur. For this reason the screen or plate winding is wound bifilar with the cathode winding. This means that the turns are put on side by side at the same time. As the full B plus voltage thus exists between adjacent turns, this method of winding requires extremely good insulation on the wire covering.

In this case the grid swing needed will be half the total output swing, or 140 volts, plus the 22.5 volts grid swing, a total of 162.5 volts, or 325 volts peak-to-peak for each tube. This is rather less than is required for the cathode-follower triode arrangement.

Note that the plate and screen must get the same supply voltage in this case because they use the same winding. This will modify the figures given slightly because these were taken for a screen voltage of 270 with a plate voltage of 360.

Distortion will be a little less than 0.3% and the source resistance about 0.13 times the load resistance.

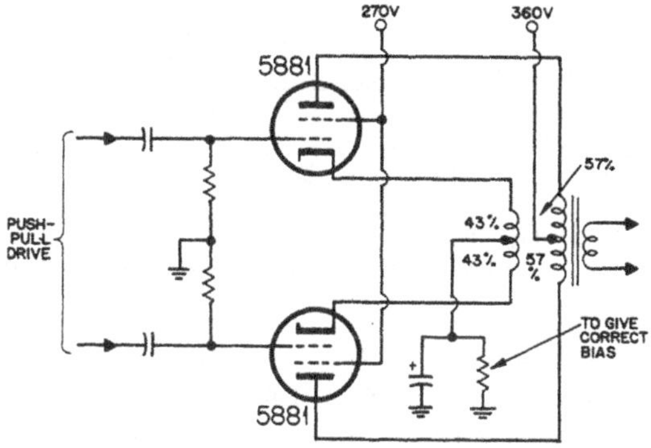

Fig. 216. *A modified ultra-linear circuit used by several amplifiers.*

Modified ultra-linear

Another variant that can be considered as a partial cathode-follower arrangement connects the B plus directly to the screen in what is virtually an ultra-linear circuit. This is shown in Fig. 216. Here to apply the ultra-linear condition strictly, the swing on the plate and cathode will be unequal so that the cathode-to-screen and screen-to-plate swing voltages are in the ratio of 43 to 57. In practice, different proportions may also be used.

This circuit conveniently enables the screens to be operated at a lower B-plus point where the tubes are better suited for operation that way. Assuming the same 400-volt supply for both, we can deduce the relative performance from the ultra-linear data already given.

The power output will be 28 watts, as before. The distortion reduces to about 0.85%, almost pure third harmonic and the source resistance to about 0.32 times the load resistance. The grid drive for this arrangement will be 45 volts grid-to-cathode, plus 43% of the original ultra-linear plate swing, 300 volts, from cathode to ground. This is 174 volts peak, or 348 volts peak-to-peak. This circuit will be recognized as forming the basis of quite a number of commercial amplifiers.

Circlotron

The last circuit we shall consider in this chapter is known as the Circlotron, shown in skeleton form in Fig. 217. This requires two separate B-plus supplies, which are shown for convenience as batteries in Fig. 217.

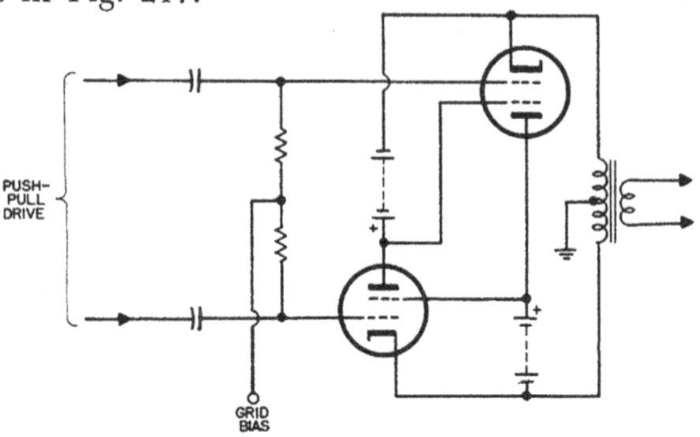

Fig. 217. *Basic Circlotron circuit. The drive arrangement is omitted for clarity in tracing the behavior of the output circuit itself.*

Why would a designer go to the trouble of setting up two completely separate B-plus supplies for an output circuit? The principal reason is that this arrangement avoids one of the problems present with all the other circuits: very special attention to the design of the output transformer. This circuit avoids the need for critical coupling between sections of the primary of the output transformer, a feature in the design of all the other special circuits. Hence a much lower cost output transformer can be used with this arrangement. This may well offset the extra cost of providing separate high-voltage secondaries on the power transformer.

These tubes are acting as pentodes because the cathode and screen are separated in each case by a constant potential. The tubes are virtually parallel, each cathode being connected to the other plate through a B-plus supply, with the output transformer being connected across the whole combination. This is where this circuit differs from the normal push-pull arrangement in which the two tubes virtually feed the load in series, not in parallel. The result is that the plate-to-plate load—or cathode-to-cathode load, whichever you prefer to call it—has a value one-quarter that used in the case of the normal push-pull pentode output.

As this is a pentode circuit, using the same operating conditions, the primary loading impedance would be 1650 ohms.

The two B-plus supplies are both "floating," one at each end of the output transformer winding. Ground is provided by the center tap of the output transformer. If the output tube grid drive was returned to ground, the arrangement would be somewhat similar to a cathode-follower circuit because the swing at the grid would

Table 2-1. COMPARATIVE QUALITIES OF OUTPUT CIRCUITS (Based on 5881 tube)

Type of Circuit	Plate Efficiency* (percent)	Distortion Amount (percent)	Distortion Dominant Harmonics	Susceptibility to Load Variation: Resistance Low	Resistance High	Reactance Low	Reactance High	Grid Drive Peak-to-Peak (volts)	Effective Source Resistance	Special attention needed
Single-ended triode	7	5.5	2nd	fair	very good	poor	good	40	0.375	Output transformer carries dc
Single-ended tetrode or pentode	43	13	3rd, 5th, 7th	fair	poor	poor	poor	36	11.5	Output transformer carries dc
Push-pull triode, fixed bics	25.5	4.4	3rd	good	very good	good	very good	90	1.4	Well-regulated B-plus supply
Push-pull triode, automatic bias	16	5	3rd	fair	very good	fair	very good	70	0.4	Accuracy of bias resistor
Push-pull tetrode or pentode	67	2	3rd, 7th	good	poor	fair	poor	45	5 (approx)	None
Simple ultra-linear	54	3.3	3rd	good	good	good	good	90	1.25	Very well-coupled output transformer
Triodes, push-pull cathode follower	25.5	1	3rd	good	very good	good	very good	414	0.31	None
Tetrodes, push-pull cathode follower	67	0.15	3rd, 7th	good	poor	fair	poor	605	.068	Bifilar-wound output transformer
Ultra-linear, cathode follower	54	0.5	3rd	very good	very good	very good	very good	690	0.17	Very well-coupled output transformer
Unity coupling	45	0.3	3rd, 7th	good	poor	fair	poor	325	0.13	Bifilar-wound output transformer
Modified ultra-linear	54	0.85	3rd	very good	very good	very good	very good	348	0.32	Reasonably good coupling in transformer
Circlotron	45	0.3	3rd, 7th	good	poor	fair	poor	325	0.13	Special B-plus supplies and drive circuit

* Plate efficiency = audio output/plate dissipation.

have to provide the grid swing in addition to the output swing on the cathodes or plates.

This is partially offset in the Circlotron circuit by returning the B-plus for each tube of the push-pull drive stage to the positive voltage from the opposite output tube. See Fig. 426, chapter 4.

Using ground as a reference, the plate and cathode will each swing a peak of 140 volts, as considered in the unity-coupled arrangement. The grid will need to follow the cathode swing, with an additional 22.5 volts making a total of 162.5 volts—still the same as the unity-coupled arrangement. This is a peak-to-peak of 330 volts but, with the cross-connection used in the plate supply, the top end of the plate coupling resistor swings 280 volts while its bottom end swings 330 volts.

This means the dynamic load line for the drive tube is multiplied by a factor of 7.2. For this reason a fairly low-value coupling resistor can be used to keep the plate voltage up but the dynamic line will be over 7 times this value. This feature enables a comparatively small drive stage tube to be used.

The power output from this stage will be the same as from push-pull pentodes with the same operating voltages. The distortion will be reduced by a factor similar to the unity-coupled arrangement and the source resistance will also be similar to it. To verify this, note that, although the tubes are connected in parallel (which reduces their resultant resistance) this reduces the step-down available in the output transformer to produce correct matching because the tubes also share the load in parallel, the center tap being provided in this circuit only to obtain a ground reference for the rest of the amplifier. So we end up with the same relationship as before.

The principal difference between unity coupling and the Circlotron circuit comes in the matter of component design. The unity-coupled arrangement requires a bifilar-wound output transformer and no particular attention to the power supply design. The Circlotron circuit can use a much less expensive output transformer but requires a special power supply with two separate floating B-plus supplies.

Table 2-1 summarizes the properties of the various circuits discussed in this chapter.

chapter 3

feedback and damping

THIS chapter will only discuss feedback as applied to power output circuits; other chapters will deal with feedback where it is used for changing the characteristics of input circuits, for use as a cathode follower in the output of a preamplifier and for equalization circuits. However, we summarize here the basic theory of feedback in relation to its use for output circuits. Since much of this will apply to other circuits, it will not be repeated—only the additional theory necessary for the particular application will be given.

Loop gain and feedback factor

In keeping with the purpose of this book, we shall not give a detailed or rigorous presentation of feedback theory, merely a summary which will permit assessing the relative advantages and disadvantages of different feedback circuits. Taking the basic feedback amplifier (Fig. 301) we have an amplifying portion with a gain indicated by the symbol A.

Some books, as well as articles, prefer to use the symbol μ to represent the gain of an amplifier. However, this author and the publishers prefer the symbol A to avoid confusion with μ as meaning the amplification factor of a tube. In dealing with feedback over only one stage, use of a symbol with two meanings would be particularly confusing because the gain of the stage is never equal to the amplification factor of the tube used in it. For this reason the symbol A is consistently employed to mean the gain, either of a single stage or of a whole amplifier.

The amplifier will have an output A times the input. For example if the input (e_g) is 0.1 volt and A has a value of 180, the output (e_o) will be 18 volts. As well as feeding whatever apparatus the amplifier has to power, this ouput is also returned through a feedback network. The voltage fed back to the input through the feedback network depends upon the feedback fraction β. In the example we have chosen, assume that 1/20 of the output voltage is fed back; that is, 1/20 of 18 volts or 0.9 volt. The fraction β is 1/20 or .05.

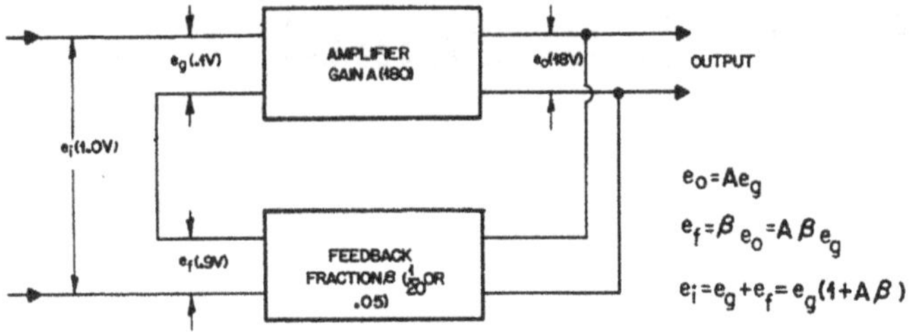

Fig. 301. *Block diagram of amplifier employing overall feedback, showing relevant quantities and formulas, and effect on overall gain.*

We have, then, a "loop gain," as it is termed, through the amplifier and back through the feedback network of $A\beta$, or, in the example chosen, $180 \times .05 = 9$. This agrees with the fact that the 0.1 volt-input gives a fed-back signal (e_f) of 0.9 volt.

This must be negative feedback, to avoid producing oscillation, so the 0.9 volt must be out of phase with the original 0.1-volt input. For a signal of 0.1 volt at the amplifier input, there is a fedback voltage of 0.9 volt so the preamplifier feeding it must supply, to overcome the 0.9-volt *negative* feedback, a total of 1 volt (e_i).

In terms of the circuit parameters, the input required from the preamplifier must be $(1 + A\beta)$ times the original input without feedback. Increasing the input by the factor $(1 + A\beta)$ to get the same output is the same as reducing the gain of the amplifier by this factor. This figure is what is termed the "feedback factor."

In the example we used, A was 180 and β was 1/20 or .05, so $A\beta = 180 \times .05 = 9$ and $1 + A\beta = 1 + 9 = 10$. The loop gain is 9 and the feedback factor is 10. To summarize:

Loop gain is the gain of the amplifier without feedback (A) multiplied by the feedback fraction (β). In Fig. 301 it is the gain that would measure putting an input in at e_g and measuring the output at e_f.

Feedback factor is the amount by which the gain of the amplifier is changed by adding feedback or closing the feedback circuit.

In Fig. 301 it is the relation between the input without feedback (e_g) and that with feedback (e_i) to get the same output e_o.

In amplifiers with the usual amount of loop gain, the two quantities are similar—they differ by only 1 and in a large number this will not seem much—so sometimes the terms are confused.

These facts have been explained in detail to clarify the distinction between the expressions "loop gain" and "feedback factor." Sometimes the figures are given in numerical quantities, in which case A is *multiplied by* β to get the loop gain. Sometimes A may be given in db and β also in db. A gain of 180, in the example noted, could be referred to as 45-db gain. The feedback fraction of 20 to 1 is an attenuation of 26 db so the loop gain is 45 – 26, or 19 db, which agrees with the factor of 9 to 1 found by simple multiplication.

The loop gain in this example would be 19 db and the feedback factor 10, which may also be referred to as 20-db feedback. The only simple way to obtain the relationship between loop gain and feedback factor is to use the numbers, rather than the db figures, which can add to the possible confusion.

Negative and positive feedback

The examples quoted refer to *negative* feedback. It is also possible, to a limited extent, to use *positive* feedback. *The limit is that the loop-gain product* Aβ *must be less than* 1. There are other practical limitations as we shall see further on.

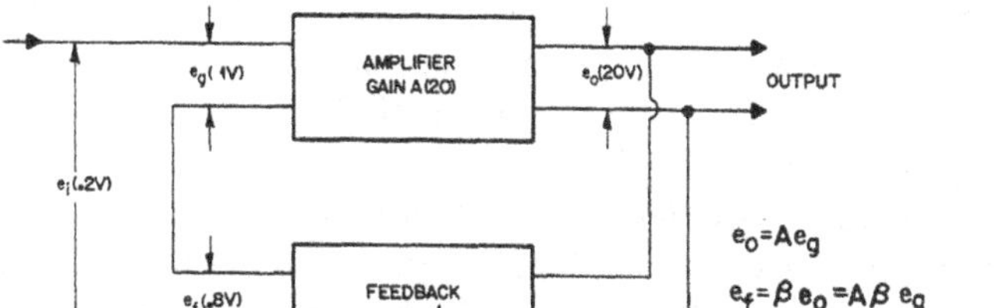

Fig. 302. *Feedback can be positive as well as negative, increasing instead of reducing gain.*

Suppose the amplifier had a gain A of 20, or 26 db, and the feedback factor β is 1/25, or .04, which is 28 db; then the loop gain will be 20/25, or 0.8, which is –2 db (Fig. 302). If the amplifier input is 1 volt, the fed-back voltage will be 0.8 volt, in phase with the original 1 volt, the signal required from a preamplifier or external source need be only 0.2 volt. The gain now is modified by the factor $(1 - A\beta)$ or 0.2.

In this example the loop gain is –2 db whereas the feedback factor is a positive one of 14 db, or a ratio of 5 to 1.

The effect of feedback on the gain of an amplifier has been given first because all the other effects that feedback produces are in some way related to its effect on gain.

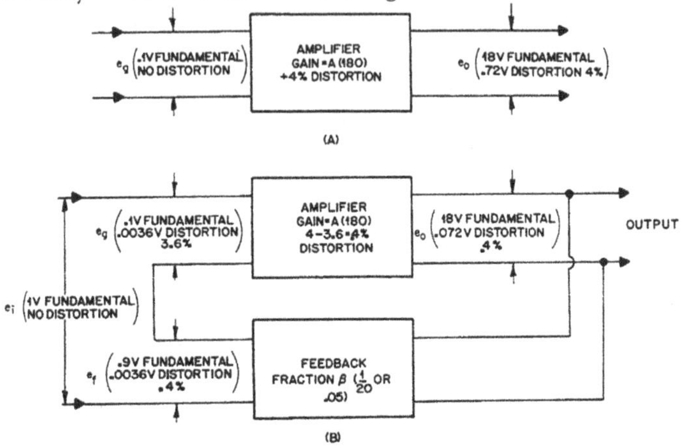

Fig. 303. *Effect of negative feedback on harmonic distortion.*

Effects on distortion

Assume that an amplifier has 4% distortion without feedback, a gain of 180 and an input of 0.1 volt. See Fig. 303. This means that, along with the 18 volts output, there will be a component –4% of 18 volts, or 0.72 volt–of distortion. This distortion may be harmonic, intermodulation or what-have-you, but it will be related to the signal. If the signal drops to 16 volts, the distortion will be 4% of 16, or 0.64 volt.

Look at the case with feedback. Theory tells us that feedback reduces distortion by the feedback factor–the same amount as it reduces gain. To verify this, fill in the figures and see how it works out. Our 20 db of feedback cuts the gain by 10:1–an input of 1 volt was needed, instead of 0.1 volt, to get the 18 volts output. So the distortion should drop from 0.72 to .072 volt (or from 4% to 0.4%). The feedback will consist of a fraction of the distortion as well as the original signal, so we shall have a feedback voltage consisting of 0.9 volt of original signal and 0.0036 volt of distortion.

The 0.9-volt original signal is out of phase with the original 0.1 volt amplified, making an input of 1 volt necessary. But the

0.0036-volt distortion had no counterpart in the original signal so it will just be fed into the amplifier, along with the 0.1 volt, to be amplified. This will multiply it by 180, like any other signal the amplifier gets, giving 0.648 volt at the output. But this is in opposite phase to the original distortion component produced within the amplifier, which would have been 0.72 volt, so we are left with the difference of $0.72 - 0.648 = .072$ volt, which is the figure we assumed.

Weakness in theory

Although this is the usual way of stating the case with regard to distortion and feedback, there is something incompatible about it—for many purposes quite unimportant—but which does account for the discrepancy and, in fact, the breakdown in the application of this theory *under some circumstances.*

When we assume that the amplifier introduces distortion, *this must mean* that the gain of the amplifier changes during different parts of the waveform. An amplifier that maintains constant gain throughout the entire waveform being amplified *cannot introduce distortion.* This means that our use of a simple number, or algebraic system, to represent the gain of an amplifier is not strictly accurate.

In the example we assumed the gain to be 180. However, this must vary by *at least* 4%, from one part of the waveform to another, if it is going to produce a distortion figure of 4%. A place where this principle interferes with the theory just developed is at the maximum output of a modern amplifier, usually limited by running into grid current clipping on the output stage.

When the output-stage grids start to conduct current, this is such a serious load on the previous stage that the voltage ceases to rise. Thus the signal at the grids is squared off at the voltage where the grid current starts and the voltage amplified beyond this point continues to be squared off. Hence there is a definite point at which the voltage excursion stops on the output waveform and, correspondingly, on the waveform passed back through the feedback.

Suppose that the input to this feedback amplifier is increased by 10% from the clipping point and that the clipping point occurs right at the 18-volt output (for simplicity we will assume that they are peak voltages rather than rms). When the input voltage reaches a value of 1 (and up to this point), feedback will reduce distortion and gain by the factor of 10 to 1 so at this point on the waveform there will be 0.1 volt across the input to the amplifier and 0.9 volt fed-back voltage to make up the combined input of 1 volt.

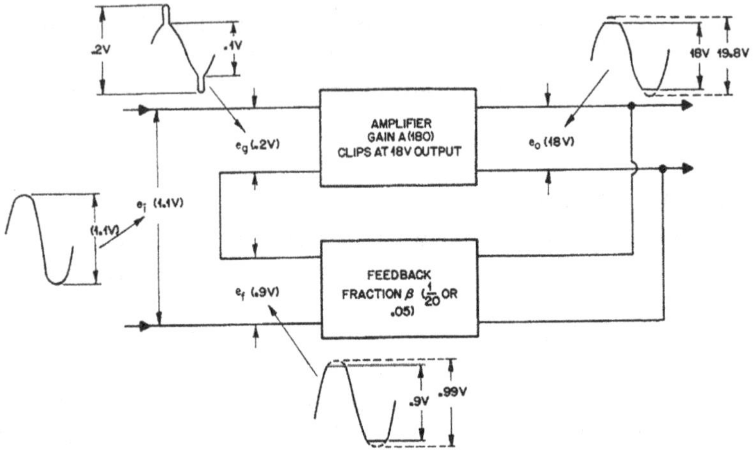

Fig. 304. *Effect of feedback on distortion is strictly limited, particularly when clipping occurs.*

But now further amplification ceases: the fed-back waveform continues at the 0.9-volt level until the input waveform gets back to the 1-volt point at the end of the clipping period. So while the input waveform goes from 1 to 1.1 volts, the feedback stays at 0.9 volt and the amplifier gets the difference, which changes from 0.1 to 0.2 volt during this interval. An increase of 10% on the input waveform is accompanied by a peak that goes 100% higher at the amplifier input. The effect of this on the waveform is illustrated in Fig. 304.

Here we see a serious implication in the effects of negative feedback when an amplifier reaches its maximum output. For an increase of less than 1 db in the input signal, the level handled by the earlier stages of the amplifier rises by as much as 6 db. In the same amplifier an increase in input of 2 db, or 26% will step up the peak input voltage from 0.1 to 0.36, or more than 10 db higher.

Cyclic blocking

Provided the stages in the amplifier ahead of the final output can handle this increased level, there is no adverse effect except that the clipping is that much more sudden than it would be with a non-feedback amplifier. In other words, the feedback cleans up the waveform right up to the clipping point and then the clipping action occurs that much more suddenly because of the negative feedback.

But, if the earlier stages are not equipped to handle this sudden increase in level, further disadvantages can occur. Suppose

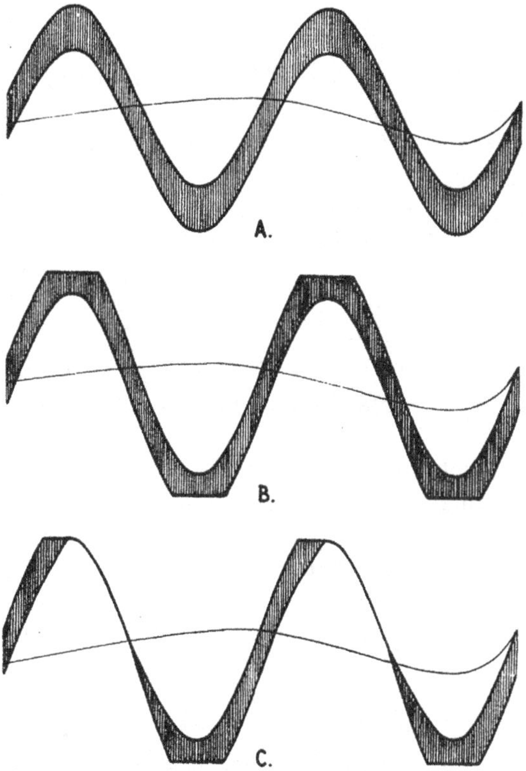

Fig. 305. *Some effects of clipping demonstrated with the test waveform used for IM testing.* a—*True composite waveform.* b—*Effect of clipping in a well-designed amplifier.* c—*Waveform that occurs due to cyclic blocking.*

that these sudden peaks cause grid current in one of the earlier stages which, due to the high circuit resistance there, alters the bias on this stage. When clipping has passed, the gain of the amplifier is liable to be changed—possibly even blocked, due to the change in bias.

This effect can be demonstrated with the aid of the kind of waveform used for IM testing, which consists of a 60-cycle wave with a 2,000-cycle wave superimposed on it. The waveform at the input is shown in Fig. 305-a. An amplifier operating without IM distortion will produce the same waveform at the output. When clipping commences under normal conditions, the waveform will be modified as shown in Fig. 305-b. But if the feedback signal causes some secondary blocking action, the waveform at the output is more like that shown in Fig. 305-c.

Avoiding overload blocking

Where an amplifier is poorly designed in this respect, a waveform of this type can occur when the input exceeds the maximum input permissible without distortion by as little as 1 or 2 db. How can this be overcome?

In the average amplifier it is not feasible to make the first stage the one that overloads first with this increased peak signal from the feedback. Due to the small signals normally handled—in the region of 0.1 volt in the example quoted—there is bound to be a considerable margin, probably as much as 10 times, before serious distortion occurs in the first stage. The stage that is far more likely to run into distortion is the second stage of a three-or-more-stage amplifier.

This means that the second tube will be likely to run into grid current and produce erratic negative biases as soon as clipping occurs in the output stage. This is inevitable. The only thing that can be done is to arrange that it will not produce the kind of after effect shown in Fig. 305-c.

The simplest way to achieve this is to have the coupling between this and the first stage of the direct variety—in other words, no time constant due to the coupling capacitor. Then, although the second stage will still run into grid current which will drastically modify the waveform at this point, as soon as the grid current ceases the voltage on the grid will be back to its correct operating point instead of having become seriously negatively biased.

This feature of negative feedback amplifier design is an unsuspected recommendation for the long-tailed type of inverter. Elimination of this particular coupling capacitor can also give advantages in attaining a satisfactory low-frequency stability margin.

Other effects of feedback

Negative feedback also reduces spurious signals, such as hum and noise, by the same ratio that it reduces gain, provided these signals are generated *inside* the feedback loop. It cannot eliminate such signals that come in along with the input. This is not often of particular value in the output circuit. It can be of advantage when overall feedback is applied with a very-high-gain amplifier, but it is of more particular value in preamplifiers.

Stability of gain

Another important advantage of feedback is its effect not only on gain but on the *stability* of gain. Reverting to the example, in which the amplifier was assumed to have a gain of 180 without feedback: let us suppose that the gain falls by 10% due to a change in supply voltage or a defective component. This means that the gain will drop to 162, where previously it was 180.

Using our formula, A has dropped from 180 to 162, but the feedback fraction β which we add has stayed the same—1/20, or .05. So the loop gain $A\beta$ will drop from 9 to 8.1. Adding 1, the feedback factor, $1 + A\beta$ will drop from 10 to 9.1. This is the amount by which the input is reduced by feedback. Previously the 1 volt was divided by the feedback factor 10 to give 0.1 volt. Now the same 1 volt will be divided by the feedback factor 9.1 to give 0.11 volt.

This will be multiplied by the changed value of A (162) to give the new output of 17.82 volts, which is a drop from 18 volts of only 0.18 volt, or a 1% change in gain compared to 10% without feedback.

So the use of feedback, giving an original feedback factor of 10, has reduced the effect of gain changes by this same factor of 10 (Fig. 306).

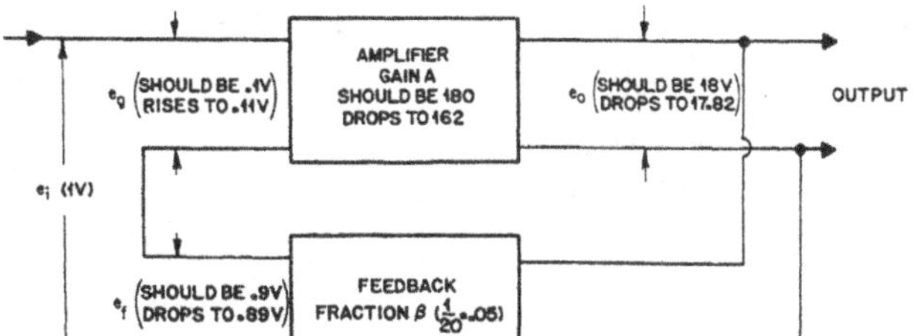

Fig. 306. *Effect of negative feedback in stabilizing the gain of an amplifier.*

Frequency response

The general impression given by many treatments on the subject of negative feedback is that the feedback improves the frequency response in the same ratio as it affects gain. Sometimes this is deduced by quite involved theory and sometimes simply by a general statement from the material presented earlier.

The claim is that if the amplifier varies in gain due to change of frequency, the feedback will produce a corresponding reduction in the gain fluctuation as that described under the heading "stability of gain." The factor now introduced and which often gets overlooked is quite a fundamental one: *any component which produces a variation in gain with frequency also introduces a considerable phase shift.*

The foregoing discussion of the relation between loop gain and feedback assumes that the fed-back signal is either in exactly opposite phase to the input signal or exactly in phase. When frequency response comes into evidence, neither of these conditions is true, hence the magnitude and phase of the fed-back signal will not agree with either of the definitions given earlier for negative and positive feedback, respectively.

There is no simple and direct method of predicting the effects of a given amount of feedback on a specific frequency response. The only way is to determine the exact frequency response of the amplifier *with* feedback in the following manner.

First, determine the frequency response, both in magnitude and phase shift throughout the loop (Fig. 307).

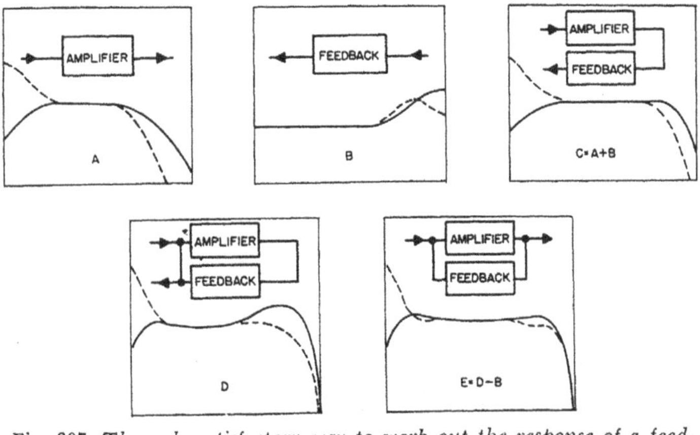

Fig. 307. *The only satisfactory way to work out the response of a feedback amplifier: First work out the amplitude (solid line) and phase (dashed line) response of the amplifier without feedback (A) and of the feedback (B), which combine to make the open-loop response (C). Then figure out the effect of closing the loop (D). Finally substract the response of the feedback portion (E).*

Second, from this information, the effect of closing the loop on the overall loop gain response can be predicted. This will, in general, depend upon how many stages of reactance elements contribute to the frequency response at each end of the spectrum and the relationship between the responses contributed by the individual elements. From this information, by appropriate calculation or the use of charts, the closed-loop response can be determined.

The final step is to subtract from the closed-loop response any response present in the feedback network. If some of the reactance elements contributing to the loop-gain frequency response are incorporated in the feedback network (contributing the function β), then the response of the forward part of the amplifier with feedback applied will be the calculated frequency response of the *closed* loop minus the frequency response of the β section (Fig. 307).

In considering the frequency response of an amplifier with feedback, a factor that must not be overlooked is the possible effect of the output load because this will invariably contribute one or more parameters to the overall open-loop response and hence also the closed-loop response.

Output impedance

Some think the use of feedback modifies the optimum load for a pair of output tubes. This basically is not true. The effect that feedback has on output impedance concerns the *source* impedance or resistance presented to the load and is used as a means of controlling what is termed "damping factor."

For triode type tubes the calculation of source resistance presented by a feedback amplifier is relatively easy, either by formula or by graphical method, using the tube characteristics (Fig. 308). First lay out the usual load line on the composite characteristics and calculate the swing voltages in all the circuit: plate volts 400 − 130 = 270; grid swing at output tubes 20; gain of amplifier "front end" 70; input to amplifier without feedback 20/70 = 0.286 volt. Fed-back signal is 1/420 of output or 270/420 = .643 volt. So the required input is 0.286 + 0.643 = 0.929 volt. This figure will be kept the same when the load is taken off. The off-load condition is found by trial and error. When the plate swing is 300 at the output (400 − 100), the grid swing needs to be 15 volts. The amplifier input without feedback would then be 15/70 = 0.214 volt and the fed-back signal 300/420 = 0.715. Then the required input will be 0.214 + 0.715 = 0.929 volt, as before. Larger and smaller swings can be tried until the right point is found.

We can now deduce the effective plate resistance. The plate swing changes by 30 volts (from 100 to 130 volts) for a change in current swing of 180 ma, representing a source resistance, referred to one half of the output transformer primary (see Chapter 2) of 166 ohms.

If the feedback is of the current type, then it will be removed by taking off the load. On the other hand, removal of the load with voltage feedback will increase the resultant feedback. Either way, the result can be computed relatively easily from the curves or calculated from the following formulas

$$R_s = \frac{R_a}{1 + A\beta} \text{ for voltage feedback;}$$

$$R_s = R_a (1 + A\beta) \text{ for current feedback.}$$

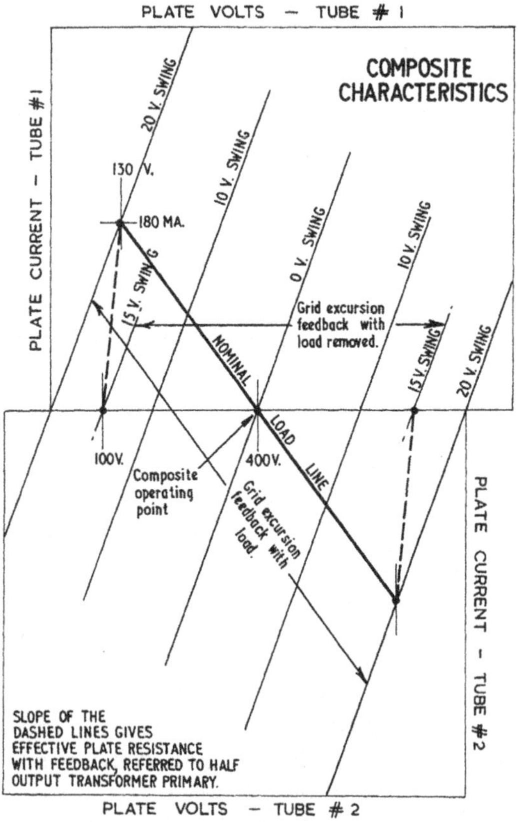

Fig. 308. *Calculation of effective source (plate) resistance characteristics.*

In each case R_s is the source resistance and R_a the plate resistance. The right values to use for $A\beta$ are discussed later. This calculation is relatively simple for the triode tube because it has a fairly constant value for all its parameters. Throughout the operating range of all load lines, the amplification factor, plate resistance and transconductance do not vary very greatly in the whole region used.

But with pentode or tetrode tubes, the matter is not quite so simple. The amplification factor of output tubes is seldom given because it carries little meaning. The most constant parameter of this type of tube is its transconductance, but both the plate resistance and amplification factor vary widely at different points on their curves. For this reason it is impossible to estimate source resistance with a feedback amplifier from the parameters given

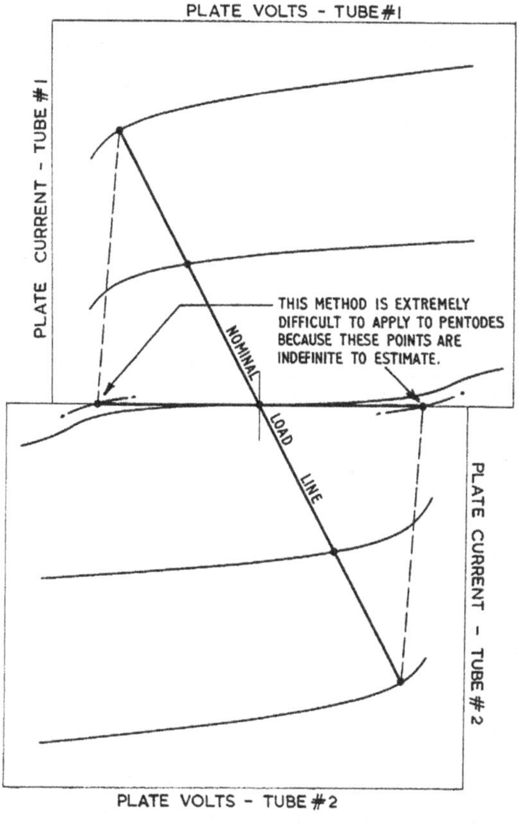

Fig. 309. *Applying the method of Fig. 308 to pentode characteristics is rendered difficult by the shallow slope and unevenness of curves.*

numerically. It is not much simpler using the characteristics, as will become evident from Fig. 309—only an approximate estimate can be achieved at best. The method is similar to that used for the triode. The difficulty is that, on the open-circuit condition, in the case of voltage feedback, the composite curves are difficult to construct because the individual curves close up and have erratic curvature. Hence it is difficult to tell where the dot and dash lines of Fig. 309 should be, and what voltage they represent. Hence the exact voltage swing obtained on open circuit is very difficult to determine. The effect of current feedback is even more difficult to judge because of the difficulty in reading the quantities involved.

QUANTITY	NO FEEDBACK		WITH FEEDBACK	
	NO LOAD	LOADED	NO LOAD	LOADED
$e_o =$	$A_o e_g$	Ae_g	$A_o e_g$	Ae_g
$e_f =$	—	—	$A_o \beta e_g$	$A \beta e_g$
$e_i =$	e_g	e_g	$(1+A_o\beta)e_g$	$(1+A\beta)e_g$
$e_o/e_i =$	A_o	A	$\frac{A_o}{1+A_o\beta}$	$\frac{A}{1+A\beta}$
$R_s/R_L =$	$\frac{A_o-A}{A}$		$\frac{A_o-A}{A} \times \frac{1}{1+A_o\beta}$	

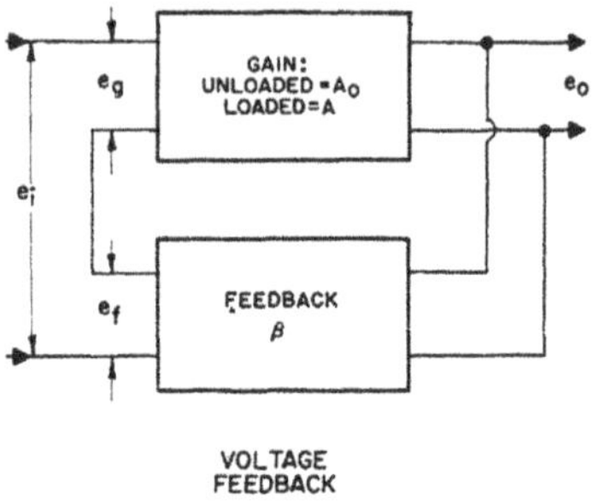

QUANTITY	NO FEEDBACK		WITH FEEDBACK	
	SHORT CIRCUIT	NORMAL LOAD	SHORT CIRCUIT	NORMAL LOAD
$i_o =$	$A_s e_g$	Ae_g	$A_s e_g$	Ae_g
$e_f =$	—	—	$A_s \beta e_g$	$A \beta e_g$
$e_i =$	e_g	e_g	$(1+A_s\beta)e_g$	$(1+A\beta)e_g$
$i_o/e_i =$	A_s	A	$\frac{A_s}{1+A_s\beta}$	$\frac{A}{1+A\beta}$
$R_s/R_L =$	$\frac{A}{A_s-A}$		$\frac{A}{A_s-A} \times (1+A_s\beta)$	

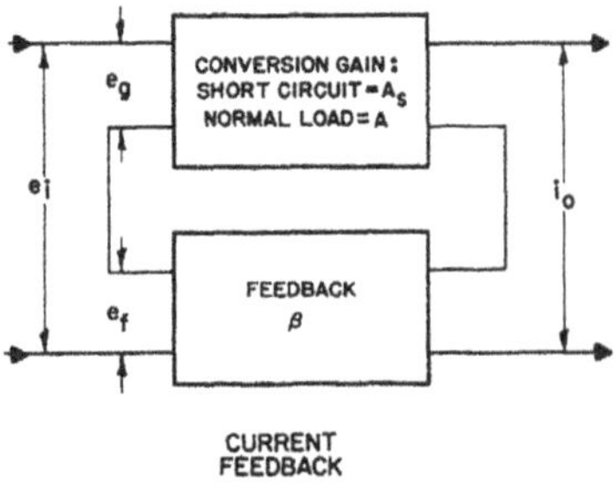

Fig. 310. *Deduction of the effect of voltage and current feedback on source resistance from first principles.*

The general effect of feedback on source impedance can be deduced in a manner similar to that used for distortion and gain stabilization, from first principles, as shown in Fig. 310. The difficulty is that the source resistance in the absence of feedback must be known. While the triode readily provides such information, the pentode or tetrode yields a source resistance which varies widely throughout the operating cycle.

To illustrate, in a triode amplifier, suppose that without feedback, the output e_o drops from 10 volts without load to 4 volts with the load connected. If the input e_i is 1 volt, A_o of Fig. 310 is 10, and A is 4, so $R_s/R_L = \frac{10-4}{4} = 1.5$. The source resistance is 1.5 times the load resistance. Now suppose that, without the load, 4 volts, or 2/5 the output voltage, is fed back. β is 2/5. The input e_i will now need to be $4 + 1 = 5$ to get 10 volts output. The feedback factor, without load $(A_o\beta + 1)$ is $4 + 1 = 5$ so the source resistance with feedback should be $1.5 \times 1/5 = 0.3$. Putting the load on should now drop the output voltage from 10 to $10/1.3 = 7.7$. We can now check this. 2/5 of 7.7 volts is fed back, or 3.08 volts, leaving 1.92 volts of the original 5 at the amplifier input. The original gain with the load connected was 4, so the output will be 4×1.92 volts, which is as near to the figure of 7.7 as calculation will give because 10/1.3 is not *exactly* 7.7.

The application of a given amount of feedback measured, in the case of voltage feedback, without the load connected and, in the case of current feedback, with a short-circuit condition, will modify the natural (no-feedback) source resistance of the amplifier by the feedback factor.

The computation of effective source resistance must be made either open-circuit or short-circuit, according to whether voltage or current feedback is considered. If the gain and feedback characteristics of the amplifier are considered with the nominal load connected, the same theoretical result can be obtained, provided the following procedure is adopted.

First, the gain of the amplifier is calculated with the load connected, hence the feedback factor is based on this calculation.

The source resistance of the amplifier is also calculated *with the load connected.* This consists of the plate resistance of the output tubes *in parallel with the load connected* for voltage feedback, or in series for current feedback.

The output resistance is then divided or multiplied, according to whether voltage or current feedback is used, by the feedback factor. The resulting impedance is a combination of the source resistance of the amplifier in parallel or in series with the load connected.

Suppose that the optimum load for the output tubes is 6,000 ohms plate to plate and each has a plate resistance of 50,000 ohms, making 100,000 ohms plate to plate. Finally, with the load connected, the feedback factor is 10 (i.e., $1 + A\beta$). The combined resistance of 6,000 and 100,000 ohms in parallel is 5,660. Voltage feedback will reduce this to an effective parallel resistance of

5,660/10 or 566 ohms. As this consists of the actual load, 6,000 ohms, in parallel with the new value of source resistance, we can deduce that the source resistance is $\frac{566 \times 6{,}000}{6{,}000 - 566} = 625$ ohms. In other words, the source resistance (625 ohms) and the load resistance (6,000 ohms) in parallel make the calculated value of 566 ohms.

The choice of method will depend upon the circuit. In pentode circuits the open-circuit gain is difficult to calculate so this second method is better. For circuits where the plate resistance and open-circuit gain are easy to calculate, the first method, using open-circuit feedback factor, is simpler and less confusing.

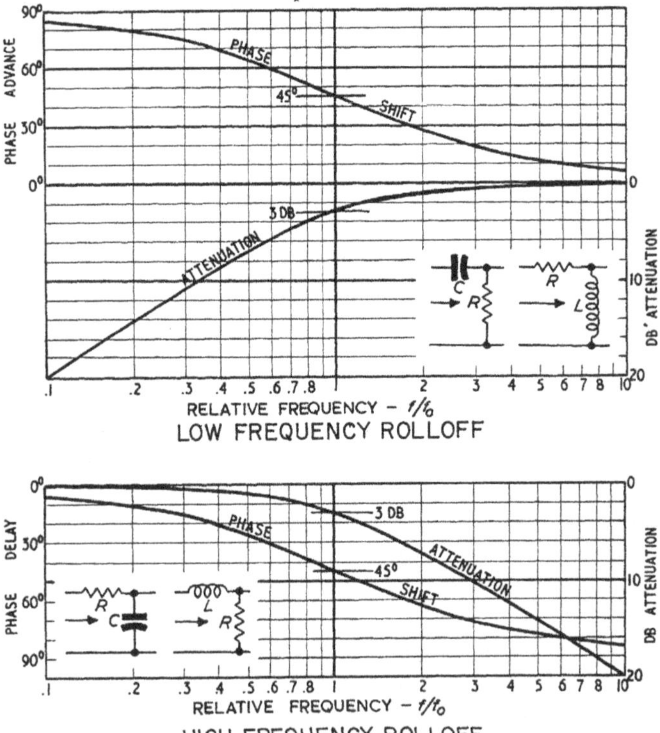

Fig. 311. *Standard rolloff characteristics for a single reactance, contributing to low- or high-frequency loss. Curves are plotted to a relative frequency in which reactance is equal to circuit resistance R.*

Stability margins

We have talked about negative and positive feedback as if they are definite entities that can be kept in convenient separate compartments. Unfortunately, they are not quite so in practice. What we design as negative feedback can become positive at some frequencies. In the formula for positive feedback, if Aβ is made equal to 1, the gain will jump to infinity, which means the amplifier will oscillate. Referring to Fig. 302, the signal fed back is equal to the original input—the amplifier supplies its own input and oscillates.

For negative feedback to become positive, become unstable, and start oscillation, it must undergo a change: its phase must reverse, or change through 180°, and there must still be at least as much signal fed back as the original input. In terms of the formula, Aβ must be 1 or greater.

Feedback does not suddenly change from negative to positive—it goes through a gradual transition. The reactances that cause ultimate phase reversal cause a progressive phase advance or delay. Reactances, such as coupling capacitors, that are responsible for low-frequency rolloff cause a phase advance. Reactances, such as stray capacitance and leakage inductance, that cause a high-frequency rolloff introduce a phase delay.

A single reactance causing a rolloff simultaneously produces a progressive phase advance or delay with changing frequency which reaches an ultimate of 90°. At one particular frequency on this rolloff an interesting relationship holds. This is where the attenuation or amplitude loss is 3 db and the phase shift is half of the ultimate, or 45°. (Fig. 311).

If in a whole amplifier there are two reactances contributing to such a rolloff and they both act at the same frequency, the loss will be a total of 6 db and the phase shift 90°. Ultimately at extreme frequencies, when there is considerable attenuation, the phase shift *approaches* 180°. In theory it never reaches 180°, only approaches it more closely with increasing attenuation.

When we move on to three reactances contributing to a rolloff at the same end of the frequency response in an amplifier, if the three reactances produce identical rolloff characteristics—that is, the 45° and 3-db frequency points are identical—then the cumulative effect will produce 9 db loss and 135° phase shift. A little farther along, each rolloff will be giving 60° phase shift and 6-db attenuation. The cumulative effect of the three networks is 18-db attenuation and 180° phase shift.

Obviously in this case, if the loop gain also happens to be 18 db at the frequency in question, we have a *positive* feedback condition with 180° phase shift and a fed-back signal equal to the input signal—just enough to cause oscillation. This is the instability point for this particular network.

Negative feedback is designed to reduce the gain (and other things that happen with it) by a large factor, such as 10 to 1. The loop gain $A\beta$ is correspondingly large for most frequencies. But, when the circuit reactances start to take effect, two things happen: there is a progressively increasing phase shift and the gain A drops off. When the change of frequency brings the phase shift up to 180°, the feedback has completely changed from negative to positive. What happens here is determined by the value of $A\beta$ at this frequency. If it is more than 1, the amplifier will go into oscillation at this frequency. If it is less than 1, the amplifier will not oscillate but the feedback will boost gain at this frequency instead of reducing it.

We can state the condition for stability in two ways: (a) when the phase shift reaches 180°, the loop gain must be less than 1 and (b) when the loop gain falls to 1, with increasing (or reducing) frequency—and until it falls to 1—the phase shift must not reach 180°.

Any amplifier with three or more reactances contributing to rolloff at the same end of the frequency response *can* reach an unstable condition if the loop gain is sufficient for the particular combination of rolloff parameters. If there is only one roll-off in the feedback loop it is impossible to produce instability, or even peaking, in the frequency response. If there are two rolloffs at the same end of the frequency response, instability can never occur, but it is possible to produce peaking.

The possibility of instability has long been recognized as an important design limitation because it makes itself noticeable so very definitely. An amplifier that goes into oscillation is unusable. The design of feedback amplifiers has long included procedure for ensuring a margin of stability.

This can be specified in two ways to correspond with the two ways of expressing the condition for stability. These are called *phase margin* and *gain margin*. Both are related to the loop-gain characteristics, and particularly the way these change with frequency. One uses the frequency where the loop gain reaches 1 (as it drops off) as a reference, while the other uses the frequency where the phase reverses (reaches 180°). They can both be calculated, but phase margin is difficult, if not impossible, to measure.

Phase margin

At the frequency where the loop gain $A\beta$ has fallen to 1—that is, the gain of the amplifier is only just equal to the feedback fraction, the phase shift must be something less than 180° to insure stability. The amount by which the phase at this frequency comes short of 180° has been called the "phase margin." This is easy to define but not quite so easy to measure.

Gain margin

The better form of margin, from the viewpoint of measurement and checking, is the gain margin. This is the amount by which the loop gain $A\beta$ is below 1 at the frequency where the phase shift is 180°. This can be checked by increasing the amount of feedback until instability occurs and then comparing the amount of feedback at which instability occurs with that actually used. (Fig. 312).

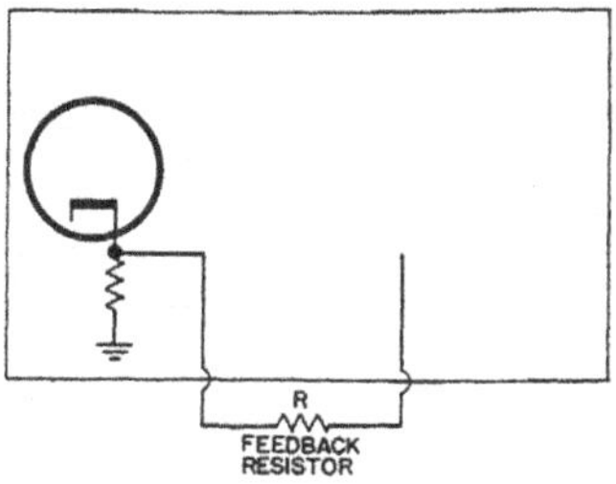

Fig. 312. *Measure the gain margin by adjusting the feedback until the amplifier oscillates.*

If the difference is, say, 6 db, then this means there is a 6-db gain margin. Many books on amplifier design give recommended figures as to both phase and gain margin for satisfactory performance. But what this method of specification does not show is the peaking that can occur in the frequency response, even though the amplifier is quite stable.

Maximal flatness

To avoid peaking at any point in the frequency response, even beyond the audio band, a margin considerably in excess of 6 db is always required. This can be particularly seen in the kind of amplifier that has only two reactances contributing to high or low rolloff. However much feedback is used, this amplifier does not become unstable; but it can run into very high peaks at the extremities of the response.

When three or more reactances for each end of the frequency response are used in the feedback loop, there is a definite margin between the point where peaking commences due to the feedback and where the amplifier runs into instability. The point where peaking commences can be regarded as the point of maximal flatness. From the viewpoint of overall performance this is the best point at which to operate an amplifier.

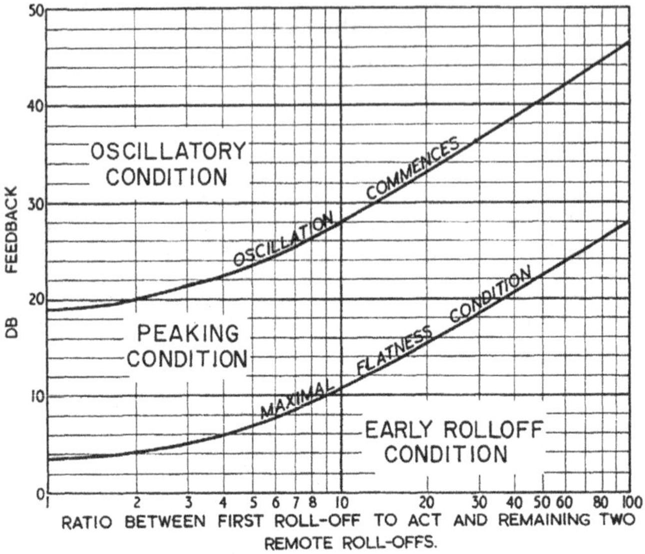

Fig. 313. *Basic relationship for various operating conditions in an amplifier with three reactances contributing to rolloff at one end of the frequency response.*

Figs. 313 to 315 show the margins necessary between the beginning of peaking and the instability point for different combinations of rolloff point in the amplifier loop-gain characteristic, using three, four and five reactances contributing to the rolloff characteristic.

Good design

The general philosophy in designing an amplifier according to this data is to have one of the rolloffs in the circuit nearer to the audio band than the rest. This allows for the maximum amount of feedback with the particular spread in rolloffs chosen.

Take the amplifier shown in Fig. 316, for example, in which the four reactances contributing to high-frequency rolloff are: (1) the plate-to-ground capacitance of the first stage (including grid-to-ground capacitance of the second stage), (2) capacitance between grids of the output tubes (with ground as a virtual center point), (3) plate-to-plate capacitance of the output tubes (including primary capacitance of transformer), (4) leakage inductance between primary and secondary of output transformer.

Fig. 314 shows that for four-stage cases 20 db of feedback requires a rolloff ratio of 55 to produce maximal flatness. A lower

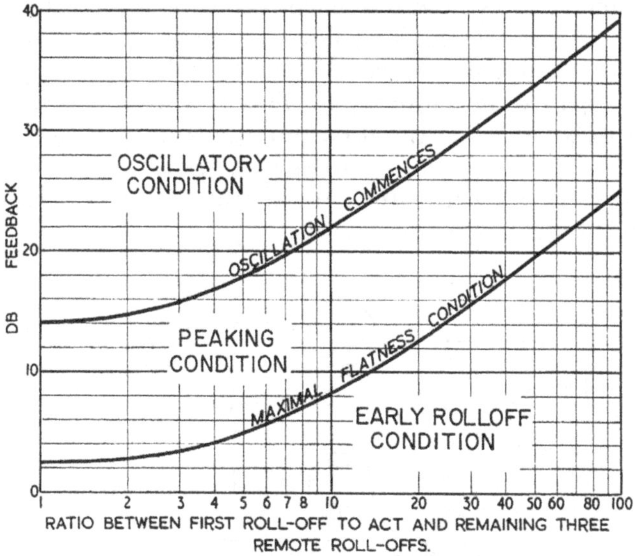

Fig. 314. *Amplifiers using four reactances contributing to low- or high-frequency loss.*

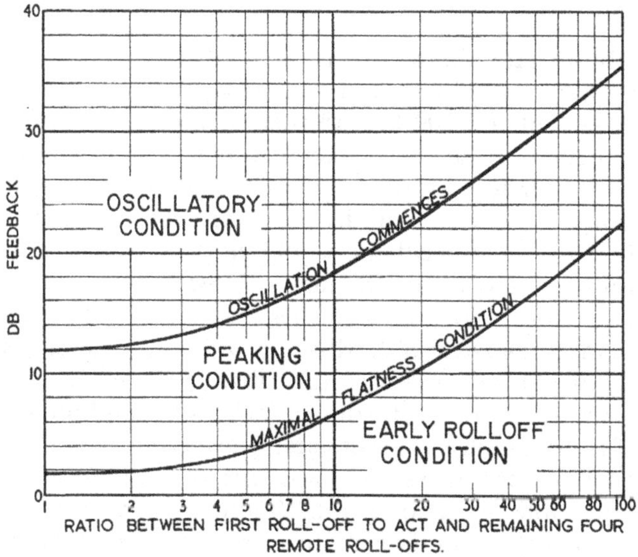

Fig. 315. *Five reactances contributing to the rolloff characteristic.*

ratio will result in a response that peaks, while a larger one (very difficult and therefore unlikely to achieve) would result in a response that rolls off earlier and more gradually than the maximal flatness curve.

For maximal flatness, the rolloff (3) could be set at 20 kc and (1), (2) and (4) at 1.1 mc. This does not mean that each has to be exactly at this upper frequency, but rather that this frequency should be regarded as a minimum. This is a difficult design to achieve, especially getting the leakage inductance that small.

Fortunately, we do not need to stretch things quite this far, because the feedback will extend the overall frequency response

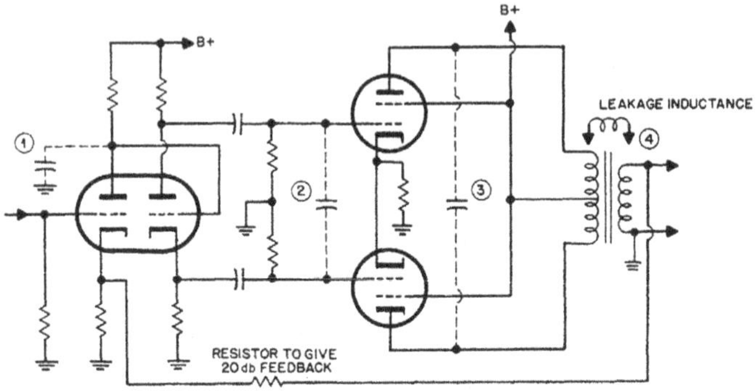

Fig. 316. *Typical feedback amplifier.*

considerably—in fact almost to the outer three rolloff points. A satisfactory design could be based on placing the upper three rolloffs at between 80 and 100 kc and then making the first one act at this frequency divided by 55—somewhere in the region from 1,500 to 1,800 cycles. This would mean that the loop gain $A\beta$ would begin to drop below 20 db at frequencies above, say, 2 kc so that the full 20-db feedback would not be available for reducing distortion above this frequency. However, the amplitude at frequencies above this never reaches maximum and so the full distortion-cancelling effect of feedback is unlikely to be necessary.

Certainly, it is never necessary to realize 20-db feedback at 20 kc because any amount of distortion at 20 kc is not audible—in fact 20 kc itself is not audible to most ears! But, taking even the highest audible frequency, its second harmonic is definitely not audible so considerable distortion of this frequency can be permitted because the distortion components are also not audible. However, the fact that the components at this frequency are very small means that little distortion will occur to need cancelling.

At the low-frequency end the problem can be a little tougher. A fair amount of power is required at least down to 100 cycles. It is not feasible to set most of the rolloffs down in the region of 1 or 2 cycles because of the oversize capacitors involved. A saving factor can be the use of direct coupling to reduce the number of reactances contributing to low frequency rolloff.

This may seem strange to many who have become used to the idea of a feedback amplifier that has a response extending way beyond 20 kc and with very low distortion within the entire audio band. To confirm the validity of our theory, let us see some of the possible ill results of more conventional amplifier design.

Super-wide-band fallacies

Very little distortion shows up over the frequency range *under test conditions*. But what happens on square waves and other kinds of transients?

Due to the fact that an amplifier so designed must inevitably possess peaking at some frequency beyond the audio band a ringing condition is set up within it. This means that, momentarily, signals of excessive amplitude at some frequency—probably in the region of 200 kc to 1 mc—will circulate around the feedback loop. This can happen although, in some amplifiers, no ringing shows up in the output. If a single test square wave is the source of this excitation, the family of frequencies set up will be interrelated and cannot produce any spurious components down in the audio band.

However, practical transients are not perfect square waves—the clash of a cymbal and various percussion sounds are not anything like the square-wave transient used for testing—and consequently contain a number of sharp exciting basic waveforms that set up a complicated pattern of frequencies in the region of the feedback peak. This means that IM distortion occurring round the feedback loop will introduce spurious components, some of which *are* within the audio band.

At the peaking frequency, the feedback has virtually disappeared as far as cancellation is concerned. It has, instead, turned into positive feedback and is almost reaching the point of oscillation. So these frequencies will be accentuated and the distortion characteristics of the amplifier, normally cut back by 20 db, are receiving full accentuated sway as far as these signals are concerned. These, then, generate IM among themselves, some components of which will be in the audio band and appear at considerable amplitude.

True, the feedback is operative at these frequencies and will reduce them by the 20-db feedback applied, but they can still be sufficient to be audible and to make reproduction harsh, gritty or muddy.

This ringing occurs even in some amplifiers that appear to give a satisfactory square-wave response. This can be demonstrated by measuring waveforms at various points (Fig. 317), which prove that the good wave at the output is obtained by skillful balance, or null adjustments, of the ringing. This does not prevent the effects just described, however.

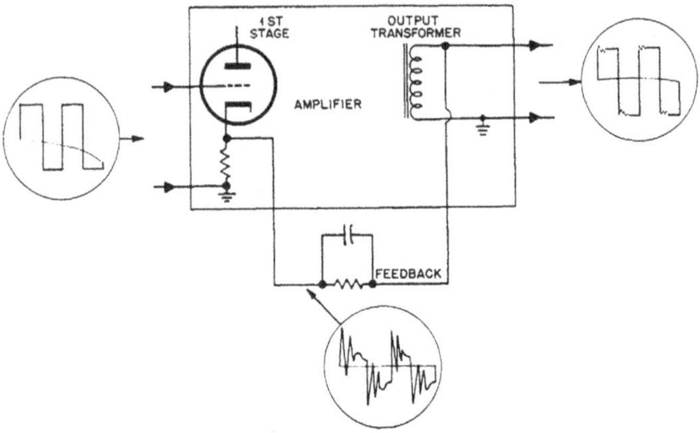

Fig. 317. *Excessive ringing, skillfully balanced in the output.*

The low end too

There is another kind of distortion that occurs regularly, due to subsonic peaking at the low end. Many feedback amplifiers may be quite stable at low frequency but, as in the case of the high-frequency response, the stability margin is probably in the region of 6 db. This means that there can be a very considerable peak at a low frequency—probably 1 or 2 cycles. This enables the amplifier to give an extraordinarily good frequency response as measured by conventional methods. It will probably be within 0.1 db, down to 20 cycles. This does not mean however, that it may not have a 10-db peak at about 1 cycle.

And how does this cause distortion? If the low-frequency peaking is increased to the point where it just causes instability, the amplifier will go into full amplitude oscillation at 1 or 2 cycles. Even on many good speakers this will not be audible because the frequency is too low to move the diaphragm at appreciable amplitude. Probably also the output transformer has low-enough inductance so that there is insufficient coupling from the output stage to drive the voice coil appreciably at this frequency.

However the oscillation will be very evident *inside* the amplifier itself. The B-plus line will be fluctuating violently and all of the tube operating points will be amplifying a large amplitude signal. This means that the gain of every stage throughout the amplifier will be fluctuating quite widely due to this large amplitude signal passing through. In turn this will cause an unusual kind of intermodulation distortion of any program material that happens to be amplified at the time.

This is one explanation of why amplifiers that give a very low measured IM distortion sometimes produce quite audible IM distortion under practical program amplification, even at low levels where distortion caused by clipping does not occur.

To complete the picture we need to know the answer to the question, "What kind of transient will start this?" Any kind that produces a "temporary dc" component in the waveform; that is, a kind of program waveform that is asymmetrical. With an old type amplifier without feedback this means that all the bias operating points will need to readjust themselves due to the asymmetry. However, in a feedback amplifier, when there is instability or peaking at this low frequency, the progressive readjustment through the amplifier will become exaggerated and sets up a low-frequency oscillation (Fig. 318).

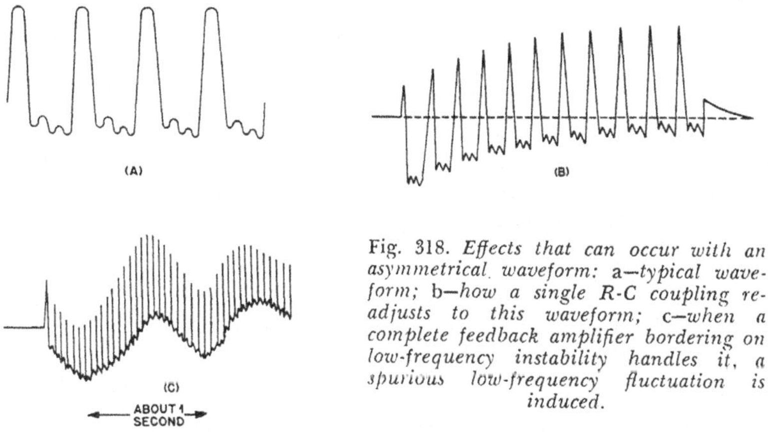

Fig. 318. *Effects that can occur with an asymmetrical waveform:* a—*typical waveform;* b—*how a single R-C coupling readjusts to this waveform;* c—*when a complete feedback amplifier bordering on low-frequency instability handles it, a spurious low-frequency fluctuation is induced.*

Typical program material that can initiate this are the plucked string instruments—particularly string bass, the drums and also many of the wind instruments, such as the trumpet. These last produce an asymmetrical waveform when viewed on the scope, probably due to the fact that there is a "dc" component caused by the wind blown through the instruments. Whatever the cause of the asymmetry, its effects can be very drastic in this kind of amplifier and cause considerable distortion even though the amplifier's specification shows it to be indistinguishable from perfect.

Loading

There is one more feature to watch in this matter of stability criteria, or margins, in amplifiers. Test conditions specify that the amplifier has a resistance load. According to specification, this means that, when the output impedance is specified at 16 ohms, the performance of the amplifier is measured with a resistance load of 16 ohms connected to it.

However, we do not listen to resistance loads; we listen to speakers and they do not have a pure resistance as an impedance. They may have an impedance of 16 ohms *at one frequency*, probably in the region of 600 cycles, but at other frequencies the impedance deviates quite widely. Down at a low frequency, in the region of the speaker's natural resonance, the impedance may well be as high as 50 ohms and in this region it will possess quite considerable components of reactance. Also at the high-frequency end there will be substantial reactance and the impedance will tend to rise.

When speakers use crossover networks, the matter is further complicated by possible reactances introduced by the crossover arrangement. All in all, the load looks very much different at all points from a true resistance load as specified for many of the amplifier tests.

Some think that if you design the amplifier with a good high damping factor, this will not matter—the feedback takes care of the amplifier and it doesn't matter what load is connected. Unfortunately, this is not necessarily true.

Assume the amplifier uses a pentode output stage in which the normal load impedance is about one-tenth of the source resistance presented by the output tubes. Using the simplified method of calculating with the load connected, a source resistance of one-tenth of the load resistance (or a damping factor of 10) can be achieved by using about 20-db feedback *when working into the correct resistance load.* This arrangement may perform perfectly but notice what happens as soon as the load changes, due to the fact that the output stage source resistance is actually 10 times that of the load resistance.

Without feedback, the voltage would rise by approximately 10 times when the load is removed. However, applying the same feedback circuit will knock the output down to the same figure it did when the load was there—within 10%. This means that the feedback has risen from its nominal value of 20 db to something probably in excess of 40 db.

If the stability criteria were calculated or measured on the basis of 20-db feedback using the correct nominal load, something very different can happen when the load is removed, allowing the feedback to rise to 40 db, and when intermediate conditions of loading are used with reactance. The amplifier has a great variety of possibilities in achieving a satisfactory stability margin. Obviously there are big—if not impossible—problems in designing such an amplifier successfully so that it will always perform equally well.

Multiple-loop feedback

The solution rests in the employment of multiple loops. Remember that feedback over only one reactance cannot cause peaking under any circumstances. So then, the best method is to apply feedback directly from the plate circuit of such an output tube to the grid or drive circuit of the same tube, to reduce the source resistance to something comparable with the nominal load resistance. This will be about 20 db of feedback. Then, having stabilized the source resistance and gain of the output stage to some extent, further feedback can be used on the overall loop to achieve a satisfactory damping factor without running into such severe complications (Fig. 319).

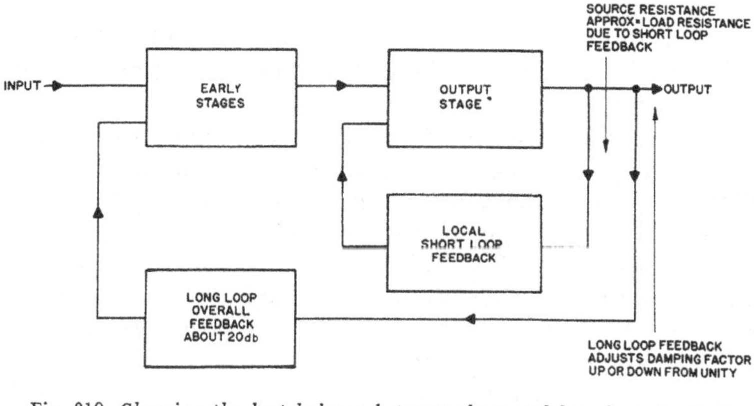

Fig. 319. *Choosing the best balance between short and long-loop feedback.*

It is appropriate here to insert a note on the relative advantages from the feedback viewpoint of different types of operation of output tubes. Triode operation yields a low source resistance, hence does not require local-loop feedback to stabilize the source resistance before applying a long-loop feedback. Ultra-linear, in effect, has a local-loop feedback in the form of connection to the screens and so achieves an objective similar to the triode. Pentode operation, however, should still use some local feedback to reduce the effective source resistance into the region of the load resistance before applying an overall feedback loop.

Phase correction methods

A common method of amplifier design consists of what might be termed "cooking" the stability margin by placing small values of capacitance in various places in the amplifier circuit. This produces what are virtually step or shelf responses in the overall loop gain. A popular position for such a "cooking" capacitor—normally given the better-sounding title of "phase-correction" capacitor—is across the feedback resistor itself (Fig. 320).

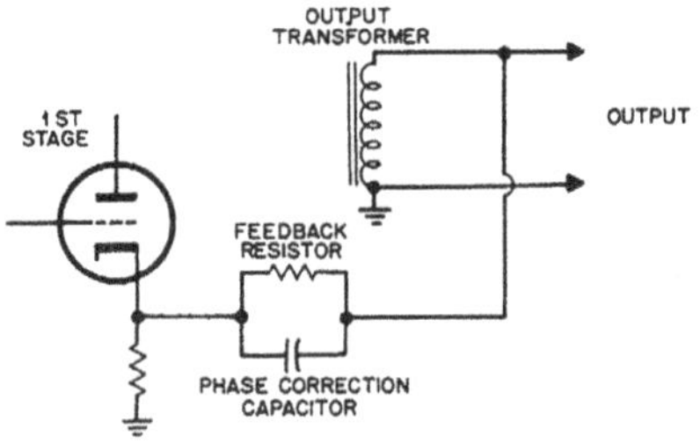

Fig. 320. *Use of so-called phase-correction capacitor in feedback amplifier design.*

The whole amplifier is carefully adjusted by means of this and other small capacitors so that the frequency response meets the specification throughout the required frequency band and so that there is what is considered to be a satisfactory stability margin in the order of 6 db.

The obvious way to see the defects of this method is to consider this design as a feedback tone-control circuit. The capacitor in parallel with the feedback resistor means that the feedback network gives a "boost" to the high frequencies. This should, in conjunction with a "perfect" amplifier, produce a rolloff in the high frequencies in the overall response of the amplifier measured from input to output. However, the amplifier has been "juggled" so that the overall response is perfectly flat. If there is a boost in the feedback section, this means that the overall loop gain, measured from input of the amplifier back to the return point from the feedback, has a high-frequency boost in it. Although the overall response, measured from input to output, may show perfectly flat, the loop-gain response inevitably has a peak because of the way in which the response is achieved.

This author has conducted some tests on amplifiers designed along these lines and revised the circuitry to fall in line with sound practice. The rolloff characteristics were adjusted to give a sufficient stability margin to insure no peaking. As a result the response of the amplifier, it is true, no longer stayed within the fine tolerances stated in the specification—which were in the region of 0.1 to 0.5 db from 20 cycles to 20 kc for different amplifiers. However, the response was still what can be considered acceptable for true high quality—within a loss of 3 db at each end of the audio-frequency response. This certainly cannot be detected audibly and, by doing it in the correct way, the amplifier sounds considerably cleaner on all kinds of program material than with the arrangement artificially extended within fine tolerances.

Multiple-loop interaction

There are some further aspects that should be considered in a little more detail (Fig. 321). Here we are considering the local-loop current feedback at the input stage in conjunction with the overall loop feedback from the output stage.

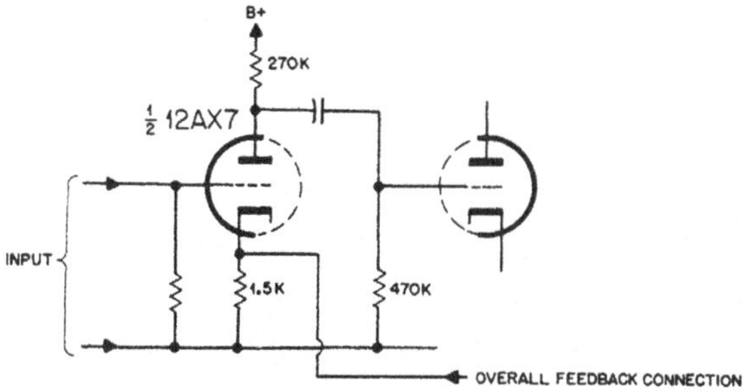

Fig. 321. *Interaction between feedback in two loops.*

Before a connection is made for the overall feedback, there is local current feedback on the input stage itself. Using the values shown, the half of a 12AX7 tube gives a gain of 67. The bias resistor of 1,500 ohms in conjunction with an effective plate load of 270,000 ohms gives a fractional feedback of 1/81. So the loop gain is 67/81 or 0.83 and the feedback factor is 1.83. This represents a feedback of 5.25 db. Assume that the amplifier is designed on the basis of 20-db overall feedback. The loop gain from input to output and back through the feedback should be 9. Now 1 volt in from grid to cathode of the first stage will produce 0.83 volt due to local feedback and 9 volts due to overall feedback. The overall reduction in gain is a ratio of 1 to 10.83, or 20.7 db. Now we see that the overall feedback is giving us 20 db whereas the local feedback has reduced from 5.25 to 0.7 db, due to the connection of the overall loop.

A similar variation happens between the loops of Fig. 319. Without a load connected, the local loop gives 20-db feedback and the overall loop 20 db more. With the correct load connected, the local loop drops to about 5 db and the overall loop to about 15 db.

"Infinite-gain" stage

The infinite gain stage is another example (Fig. 322). Infinite gain is achieved by connecting the resistor R between cathodes of

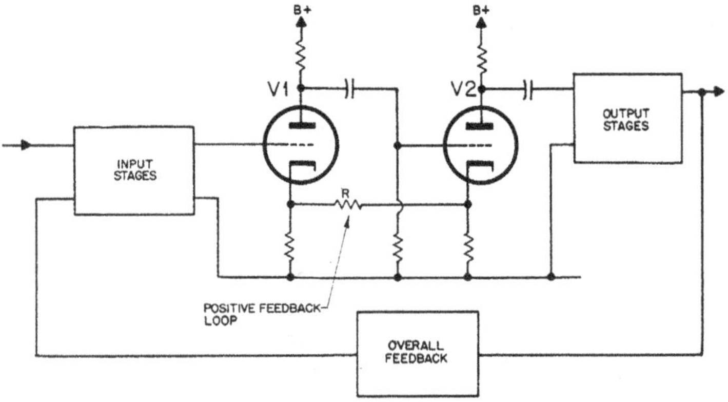

Fig. 322. *Another case illustrating interaction between feedback loops.*

the two stages shown. This means that the signal at the cathode of the first tube is equal but in opposite phase to the signal at its grid. This anti-phase voltage is derived by feedback from the cathode of the second tube shown, which cancels the current fluctuation due to the first tube itself and exceeds it to this extent.

If the grid of the first stage is considered as strapped to ground, this two-stage setup will oscillate freely at a frequency determined by the reactances of the overall arrangement. This will cause phase shift to prevent oscillation at frequencies other than the center of the band. If feedback in excess of a loop gain of 1 is employed, the oscillation will become "harder," producing a squared or clipped waveform, until the effective loop gain averaged over this distorted waveform is 1.

Assume the overall loop applies what would be 20 db if the positive feedback were not present—that is, if R were removed. As a start, assume a signal at the plate of the first stage shown that would normally be occasioned by 1 volt from grid to cathode. Due to the 20-db negative feedback there will be 9 volts in opposite phase delivered via the output, feedback loop and input—that's assuming nothing runs into distortion. This would be so wherever you start the consideration of a feedback amplifier. It can be shown at the input because 9 times the input comes back in opposite phase from the feedback network, but when the input circuit is closed, the same thing will happen if the amplifier loop is opened at any other point.

When R is connected, there is an equal voltage of 1 returned from the cathode of V2 so that the two stages would go into oscillation. But at the same time, due to the same hypothetical initiating voltage of 1 volt between grid and cathode of V1, the overall negative feedback loop will return an anti-phase voltage of 9 volts to the grid. So now the overall effect with the positive loop feedback between cathodes is as if the negative feedback loop had a loop gain of 8 instead of 9. In other words, the effect of closing both loops now has a net change on this stage of 19 db.

However, the overall feedback has been set at 20 db, compared to the condition without positive feedback. The fact that there is now only an *apparent* negative feedback (referring to this stage) of 19 db, indicates that the positive feedback, instead of being infinite as in the absence of the negative feedback loop, now merely boosts the gain of this stage by 1 db—the difference between 19 and 20.

It's all a question of viewpoint. When you consider the overall amplifier, there is still an apparent gain of infinity before the negative feedback loop is connected. So it is possible for some things to be done, using this circuit arrangement, that are not possible with a simple straightforward 20-db negative feedback. But we have to be careful how we apply this thinking.

If the infinite-gain stage produces any distortion, this is likely to get magnified. Without the negative loop feedback from output to input, the stage has infinite gain and considerable distortion (self-generated). Applying the negative loop feedback brings the gain down from infinity to some finite value and at the same time restricts the distortion in this stage—or would do so if the negative feedback had no distortion elements of its own. But, because of the distortion elements in the negative loop feedback, particularly the output stage, this fact is no longer true and the distortion elements themselves will interact in a way that becomes somewhat complicated to predict.

This can be particularly troublesome when the output stages are driven to clipping point. Then the negative feedback momentarily disappears, leaving the "infinite-gain" stage free to "take off" until clipping ceases.

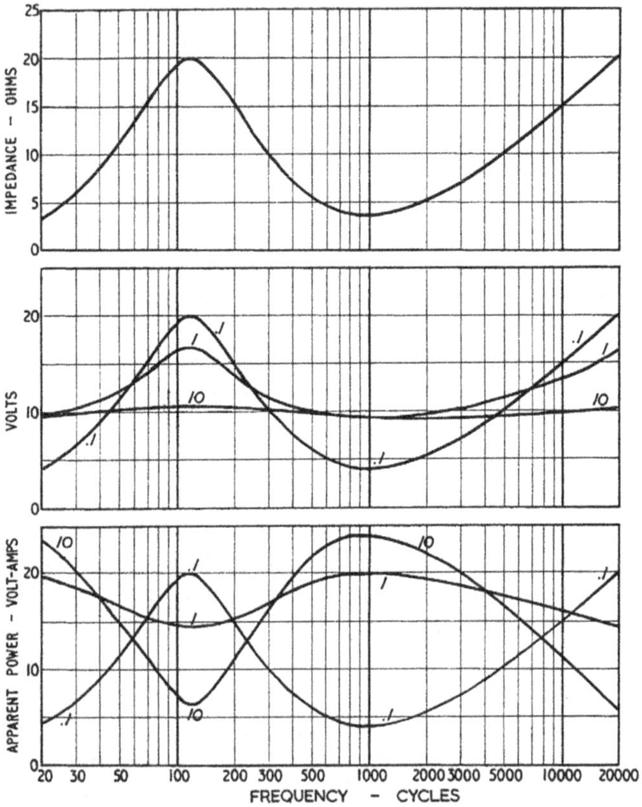

Fig. 323. *Impedance characteristic of a speaker affects the response fed to it, using different damping factors.*

Positive feedback has some limitations associated with overall stability and response criteria. Over a single stage, feedback merely modifies bandwidth—positive narrows it, negative widens it—and the overall criteria are simple resultants of both forms. If positive feedback is used over more than one stage, the bandwidth is still narrowed but the associated phase shift is no longer simple so the ultimate interaction with a negative feedback loop can produce complicated and almost unpredictable results.

Variable damping

Variable damping in itself produces some fundamental problems. If the speaker were a pure resistance, the question of damping would be quite unnecessary because a resistance is critically damped in itself. However, since a speaker possesses both electrical and mechanical or acoustical forms of reactance, damping is necessary. Unfortunately, the fact that these reactances are present

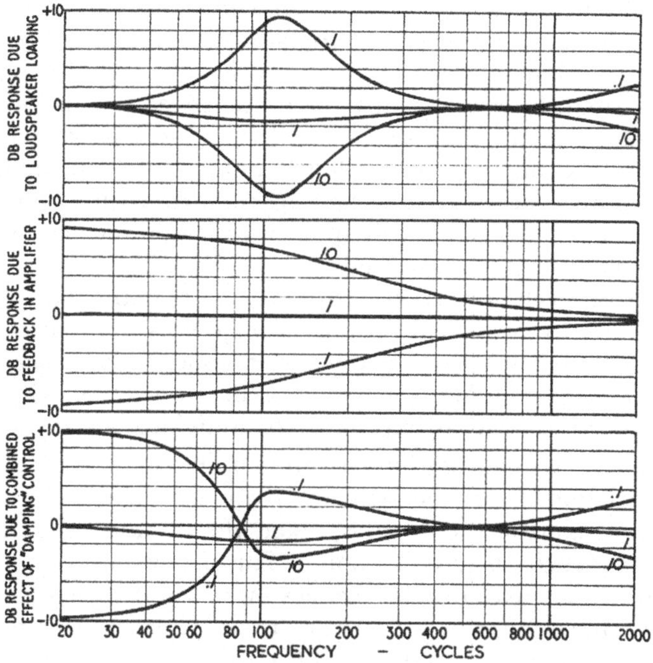

Fig. 324. *Performance of a variable damping control that uses current feedback at the low-frequency end only. Top—variation in frequency response caused by different damping factors acting with the speaker impedance characteristic (as in Fig. 323). Center—amplifier response changes as the damping factor is changed. Bottom—combined response due to both effects acting together.*

causes the speaker to have a somewhat complicated load-impedance characteristic. This means that the frequency response of the energy fed to such a load will vary according to the resistance source from which it is fed.

If the source resistance is zero, the voltage across the load will be constant at all frequencies. As soon as the source resistance is increased to some finite value, the voltage across the load will follow the general contour of the impedance characteristic (Fig. 323).

Variable damping aims to eliminate undesirable resonant effects from the speaker by damping transient vibrations. This is not directly connected with the frequency response of the instrument and should be considered as a completely separate form of distortion. However, as soon as you start to adjust the resistance from which the loudspeaker is fed, you will also alter the frequency characteristic it gives.

Some of the earlier damping-factor control circuits compensated for this by arranging that the adjustment in feedback occurred only at low frequencies. This is where control of damping is most desired. By varying only the current feedback at low frequencies and keeping the component of voltage-feedback constant, the gain of the amplifier at the low-frequency end was modified by the damping factor in a manner approximately inverse to the effect of this control on the frequency response due to the impedance characteristic of the loudspeaker (Fig. 324).

Unfortunately however, in designing the amplifier, the compensation has to be judged for the *average* speaker; that is, the change in current feedback takes over at a specific frequency designed into the amplifier. This may or may not agree with the position of the change in impedance of individual speakers.

Although this compensation does to some extent level off the response variation *above* the resonant frequency in the electrical-impedance characteristic of the speaker, the variation is exaggerated at frequencies below this point. This means that going into the region of positive current feedback—which is the maximum or negative damping-factor region—the response is exaggerated at points below speaker resonance. This will include rumble frequencies and similar components from the amplification system, hence results in an undesirable exaggeration of any subsonic frequencies present. If, in addition, the amplifier has a peak in its low-frequency characteristic due to poor stability margin, the distortion introduced can become considerable although this might appear to be the optimum damping for a particular speaker.

At best, any attempt to control the damping factor must be a compromise. It seems that the change in frequency response, due to the impedance characteristic of the speaker, which inevitably accompanies adjustment of damping factor must be accepted and, if necessary, further measures should be taken to compensate for undesirable features in this characteristic.

What method shall be adopted to change the source resistance of the amplifier? There are two basic methods left. One uses combined voltage and current feedback in a variety of circuits of which Fig. 325 shows the essential features of two typical examples. They both achieve the same objective, but *b* produces less loss of power in the current dropping resistors. However *a* has the advantage of simplicity in not requiring an unusual type of two-gang potentiometer.

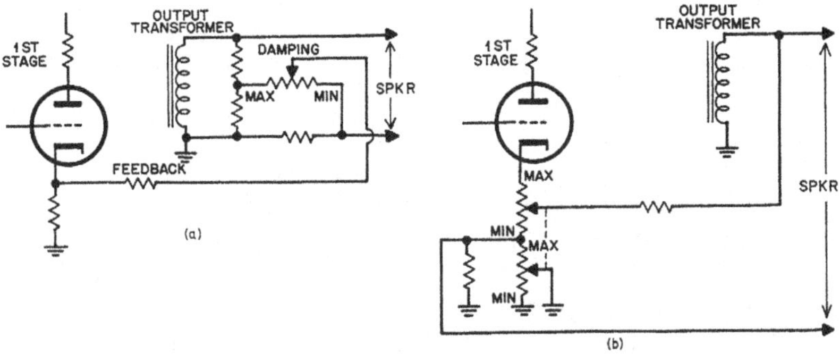

Fig. 325. *Two variations of circuit that uses constant total feedback (assuming nominal loading), but varies the proportions of voltage and current feedback to change the damping factor.*

If the amplifier without overall feedback has a damping factor in the region of unity—that is, source resistance equal to nominal load resistance, then a variation in the current and voltage feedback, which adds up at all times to a total of 20 db, will give a deviation in damping factor from 0.1 to 10. If the amplifier is correctly designed, apart from the damping-factor control, for the use

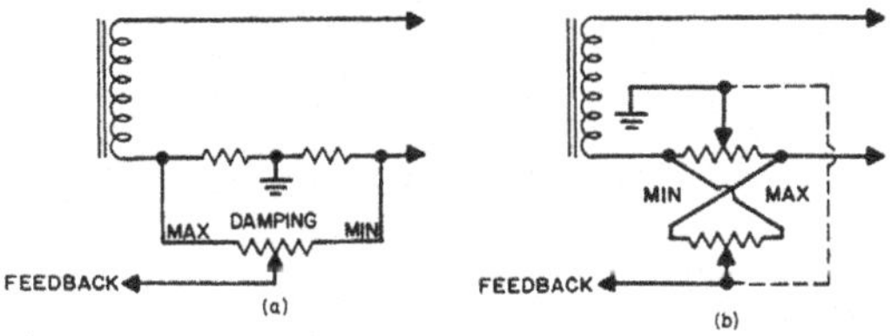

Fig. 326. *Two variations of circuit that uses positive and negative current feedback to adjust the damping factor.* a—*Uses only one potentiometer.* b—*Requires only half the series loss to obtain the feedback voltage, by using a two-gang potentiometer.*

of 20-db overall feedback, the stability criteria and frequency response should be satisfactory. Changing the proportions of feedback by use of a ganged control will give this available variation of damping factor without any other undesirable effects.

The other form of damping-factor control has the questionable advantage that the source resistance can be made negative. Working in terms of source resistance, rather than damping factor, this means that the resistance can be reduced to the zero point, corresponding with infinite damping factor, and then can be further made negative to neutralize some of the voice-coil resistance. This should improve the coupling between the amplifier and the voice coil movement because it is like winding the voice coil with a better conductor than copper or aluminum.

The only simple way to achieve this is to use positive and negative current feedback (Fig. 326). Negative current feedback reduces damping factor or increases the source resistance. Positive current feedback works to a point where the source resistance becomes zero—that is, the output voltage does not change as the load resistance is varied and then, taking the positive feedback a little further, the output resistance even goes negative—putting a load on the amplifier increases the output voltage of the amplifier.

The basic disadvantage is that the stability criteria of the amplifier as a whole cannot possibly be kept constant because the total feedback used in the overall loop cannot be constant. Although the amplifier characteristic can be kept within acceptable tolerances in the audio-frequency range, there may still be peaking at sub- or ultra-sonic frequencies with the attendant distortions that this can produce. The use of positive feedback in this manner over a number of stages also complicates the problem, as mentioned in connection with the use of the "infinite gain" stage.

chapter 4

inverter and driver stages

THE basic requirements of this section of the amplifier are (1) to provide a balanced drive signal for the grids of the output tubes from a single-ended input and (2) to provide *enough* swing of this balanced output to drive the output tubes to full power. Of course, there should be a slight margin here, at least to insure that the output tubes are the limiting factor for the maximum power delivered by the amplifier and not some earlier stage.

Factors considered

Each circuit will be fully considered in relation to the available grid swing that it will give for a specified B-plus voltage. It is recognized that the swing any tube gives is a function, not only of the tube characteristics, but of the B-plus voltage available for its plate circuit.

Next in importance is the degree of balance that can be achieved by the circuit and also how much the balance of the drive provided for the output tubes depends upon the individual tube parameters. Circuits which give a high degree of balance regardless of deviations in individual tube parameters, are obviously better than those which require careful matching or control of tube characteristics.

Not only must the balance be arranged to give accurate antiphase voltage to the grids of the two output tubes, it must also preserve this relationship accurately *throughout the frequency range* handled by the amplifier. This means that the phase inversion must be accurate at all frequencies.

Another important feature is the amount of distortion it contributes to overall performance. Earlier stages in an amplifier usually operate at relatively low levels where (if the distortion is still on the large side) it is a simple matter to provide overall voltage feedback to reduce the distortion to acceptable limits. But, like the output stage, the phase inverter and driver stages handle relatively large signals. The output stage delivers relatively large power and the phase inverter and driver stages work with relatively large voltages. So the distortion—not only as to quantity, but as to kind—is important.

A final feature is how the phase inverter and driver stages affect the use of feedback. This can be considered from two viewpoints:

1. The design of the phase inverter and/or driver stage may need to include arrangements for short-loop feedback from the output stage to provide a satisfactory starting point for raising the damping factor to a successful figure. For this the design must have convenient points to which to make the local-loop feedback connection.

2. Different circuits will contribute different frequency response parameters that will affect the overall stability criteria of the complete loop for low- and high-frequency response determination.

Direct-coupled transformers

The first and obvious type of phase inverter to be used was the simple transformer, connected directly in series with the plate in the preceding stage (Fig. 401). This component has had little use since the early days of amplifiers, largely because it was dismissed as a prominent cause of distortion. This was due to two principal factors: (a) inadequacy of core materials at that time for audio applications (modern core materials have made possible considerably improved audio transformers) and (b) the fact that most of the available transformers were nowhere near as good as they could have been made with a little more care in design.

At one time, when audio amplifiers were part of a radio receiver, quality was not so important as gain. The average audio section consisted of a number of low-magnification battery-operated triodes, coupled by high stepup interstage transformers. These had any number of turns wound on some convenient bobbin plus laminations of nondescript steel. The whole was then placed in a container to hide the shabby construction and this was sold as an audio transformer.

Both the frequency response and the distortion of these units were very poor. Long before the feedback amplifier these components had fallen into disuse, except for limited applications, because of prejudice against them arising from their poor workmanship.

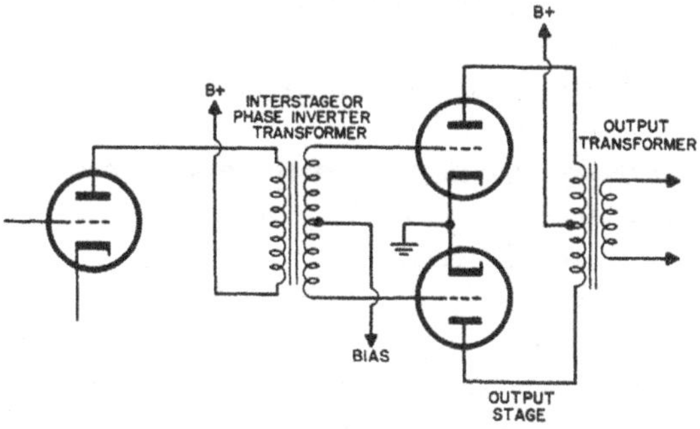

Fig. 401. *Simple circuit for transformer phase inversion.*

Yet, a well-designed interstage transformer, including one of the phase-splitting variety, can yield quite good performance. In fact when correctly designed, it has many advantages over some of the later types of phase splitters. Due to the relatively low primary resistance, the plate voltage need not be much lower than the B-plus voltage. This enables the tube to deliver a large swing on the primary side of the transformer.

In addition, the use of a low-mu triode such as half of a 12AU7 permits a considerable stepup ratio to be achieved between primary and each half-secondary. Thus this circuit will certainly beat any other for providing the maximum voltage output handling capacity for a given B-plus voltage.

It also scores on the stability of balance against tube parameters, for the reason that the balance is dependent only upon the transformer's being accurately wound in the first place. If the tube gain drops, then the output on both sides of the secondary will drop in the same proportion.

It is possible to achieve a high degree of balance in the transformer frequency response. At the low-frequency end the inductance produced by the core, reflecting to the primary winding, produces a low-frequency loss and phase shift. The stepup ratio of the transformer produces a replica of this frequency response both in amplitude and phase shift for each grid so that, whatever the frequency characteristic produced on the primary, an identical one, at larger voltage because of the stepup, will appear on both halves of the secondary.

At the high-frequency end more care in the design of the transformer is necessary to maintain balance. Nevertheless a well-designed unit can provide identical winding capacitances and leakage inductances for each half, thereby maintaining uniform phase relations between the drive for the two outputs.

Careful attention to the associated operating impedances can also achieve a much better frequency response than is usually believed.

Due to the unhappy experiences of early days, interstage transformers are widely accused of producing distortion. However, a well-designed component will result in a given voltage swing at the output with much less distortion than any of the newer tube circuits. This may come as a surprise to some readers but it has been verified both in design and practical comparison. Of course a high-quality interstage transformer is more costly than the extra tube which might be needed to replace it, with a few extra resistors to help out, but performance-wise the transformer wins *except on one point.*

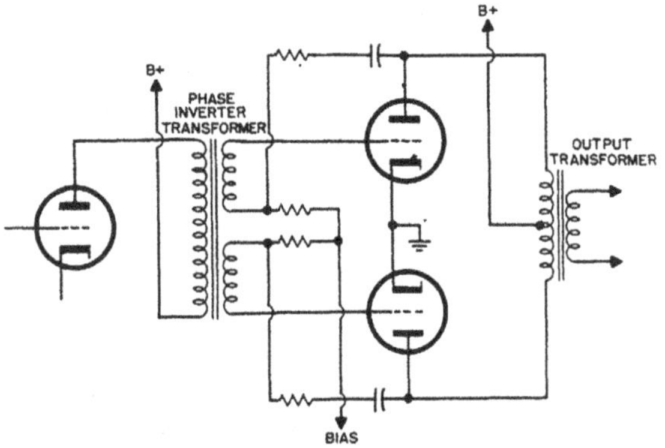

Fig. 402. *Transformer circuit provides short-loop feedback without absorbing appreciable power from the output circuit.*

This is its relation to feedback circuits. From one standpoint the interstage transformer even has an advantage here (Fig. 402). It enables a local-loop feedback to be arranged over just the output stage by itself—an arrangement which proves to be ideal because it is inherently stable and cannot produce peaking. Because the feedback resistors are virtually in the grid circuit of the output tubes, this arrangement enables values to be used which do not absorb appreciable power from the output—a factor which discriminates against the use of extensive local-loop feedback when this is connected into the cathode of the driver stage.

There is a limit to the amount of negative feedback that can be applied in this manner because there is also a limit to the swing that can be obtained from the preceding stage and to the amount of stepup that the transformer can provide with a reasonable frequency response. It will be realized here that the frequency response of the transformer does not come inside the local-loop feedback.

The real disadvantage in the use of any type of interstage transformer comes at the high-frequency end when considering overall feedback. At the low-frequency end the transformer does not have any serious advantage or disadvantage compared to resistance–capacitance coupling. R-C coupling introduces a single reactance producing low-frequency rolloff and the primary inductance of the transformer does the same thing in a different manner. So, for this purpose, both methods of coupling start level.

But at the high-frequency end the transformer contributes a number of reactances. There is the primary capacitance, the leakage inductance between primary and secondary and the secondary capacitance. This results in a network whose ultimate high-frequency phase shift is 270°. Compared with this, any R-C coupling between stages produces a maximum phase shift of only 90°. Additionally, there is no serious difficulty in designing an R-C-coupled stage whose 3-db point is as high as 100 kc, or even higher, whereas the reactances combining to effect rolloff in a transformer can be pushed up to around 20 kc but this is about the limit. This means that a single interstage transformer in an amplifier with, say, three more stages, cannot use more than about 6-db overall feedback before instability occurs. Almost any feedback at all will begin to produce peaking in such an amplifier.

This is the principal reason why it is not practical to use interstage transformers in modern amplifier design. However the facts have been presented fully so that application of interstage transformers can be considered where this reason may not be pertinent.

From the designer's viewpoint the direct-coupled transformer has problems, particularly in achieving a wide frequency response. Since current flows through the primary winding, it is difficult to achieve a high primary inductance. The core has to be gapped to reduce saturation effects. This means there is little advantage in using the high-permeability modern materials. Consequently the direct-coupled transformer is inherently limited to a somewhat restricted frequency range.

Parallel-fed transformer

This situation is considerably improved by the use of parallel coupling (Fig. 403). This avoids passing the plate current through the primary of the transformer and makes for a much larger primary inductance with fewer turns on a smaller transformer. With improved core materials, a parallel-fed transformer can attain much wider frequency range than a direct-coupled transformer

However, it has certain strikes against it immediately. Because the plate has to be fed through a coupling resistor, the working plate voltage is much lower. This will restrict the swing available on the primary to half, or maybe less, than that available for the direct-coupled transformer. So, to produce the same swing on the secondary, the parallel-fed transformer needs to give twice, or more than twice, the stepup designed into the direct-coupled transformer type. Usually this can be done.

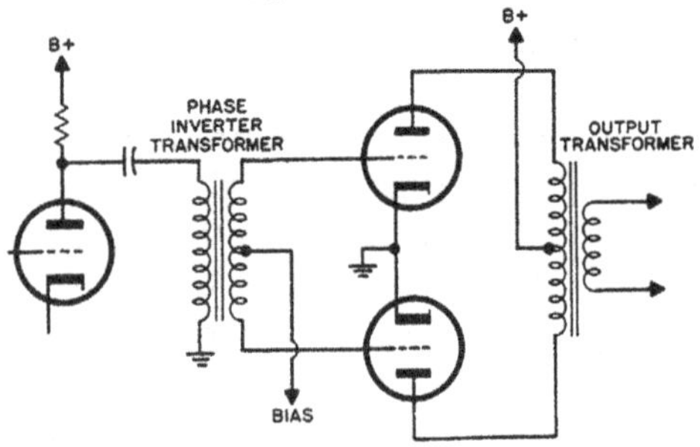

Fig. 403. *Parallel-fed transformer improves frequency band possibilities and makes a bigger stepup available.*

But there is yet a further disadvantage. At the low-frequency end we now have two reactances contributing to rolloff instead of the simple inductance of direct feed. The coupling capacitor produces a loss of low frequencies while the primary inductance provides a further shunt loss. These two components have to be correctly matched with the circuit impedances to provide a good low-frequency response without peaking and without undue loss.

At the high-frequency end also, the change in stepup introduces troubles that offset some of the gain achieved by the change in transformer type. With the much smaller physical size and bigger stepup, the primary capacitance effects are negligible compared to the direct-coupled type. Although the winding capacitances on the secondary side are very much less than those in the physically larger direct-coupled transformer (the leakage inductance can be considerably smaller on this account) the larger stepup ratio means that the *effective* source resistance provided by the plate circuit of the preceding tube is about quadrupled. This means that the leakage inductance and secondary capacitance must be divided by 4 if their effect on high-frequency response is to be of the same order as in the direct-coupled case.

Most of the advantages gained by parallel-feeding the transformer are in effect lost due to its disadvantages.

Tapped af choke

Another variation that uses inductance is the tapped af choke circuit (Fig. 404). This provides phase inversion without the stepup that the parallel-fed transformer will give. One of the output grids is directly coupled (via a capacitor) to the voltage amplifier plate while the other is inductively coupled.

Its stability of balance with tube parameter variation is as independent as any transformer. Its distortion is about the same as any voltage amplifier stage, giving the same swing *each* side as the voltage amplifier delivers. By not too complicated design the response can be made very good although the balance will always go off at some high frequency because of the leakage inductance between halves, which only affects the "lower" output grid. Also, it is not readily suitable for local-loop feedback because there is no isolation between primary and secondary, as with a full transformer.

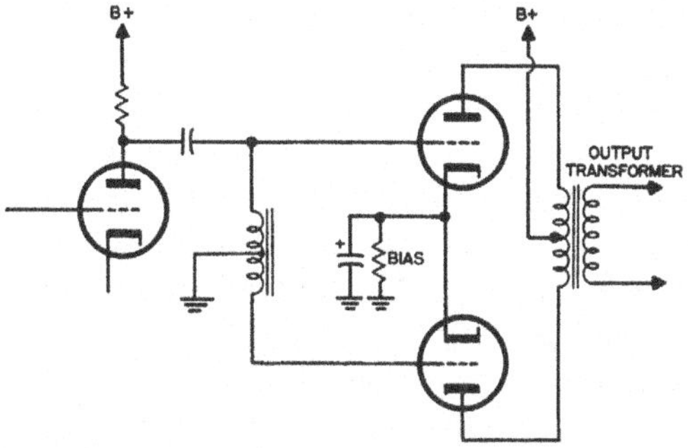

Fig. 404. *Circuit using a center-tapped audio choke.*

Split-load inverter

The first type of tube phase inverter to be introduced is called by various names. The most descriptive would seem to be "split load." However other names, such as concertina, fifty-fifty, cathodyne and the like, have also been used. The basic circuit is shown in Fig. 405. Half of the plate load is connected between the plate and B-plus while the other half is connected between cathode and ground. Since, at any instant, the same plate current flows through both these resistors, the voltages must both change accordingly. If the resistors are made identical, then both halves will be identical.

Obviously, however, the swing that can be derived from this stage will only be half, at the plate or cathode, of the swing that could be achieved using the tube with the same total resistance,

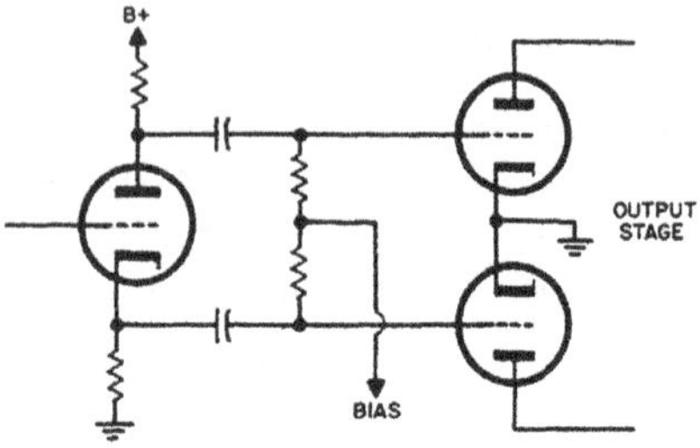

Fig. 405. *Basic circuit of split-load inverter.*

all in the plate. In addition, the previous stage has to deliver a driving voltage slightly larger than the output swing obtained.

As an example, a 12AU7, using typical values, will give a gain of about 10 into the total plate load chosen. This means 5 at the plate and 5 at the cathode; so 1-volt swing between grid and cathode will give 5 volts between cathode and ground and 5 between plate and ground. But the input swing needs to be the combined voltage of grid to cathode and cathode to ground—a total of 6. So, with the 12AU7 as an example, the input voltage has to be 1.2 times the output voltage to each grid.

The most serious limitation of the split-load phase inverter is that its output is restricted because the load is split. It may even be necessary to employ a further stage of amplification between this phase splitter and the output tubes themselves. Additionally, a large input is needed; the tube does not provide any gain—instead it gives a slight loss.

As regards balance in frequency response, the circuit is excellent. This is contrary to what many authorities write so perhaps it will be well to investigate here just what this tube does.

The reason given for describing the tube as giving poor balance is that the source resistances at the plate and cathode are very different. From the viewpoint of the plate, the tube is operating with a large degree of current feedback due to the resistor in its cathode while, from the viewpoint of the cathode, the tube is operating halfway to being a cathode follower. So the resistance "seen" at the plate is several times the ac resistance of the tube, in parallel with the actual coupling resistor used, while the resistance "seen" at the cathode is much lower than this. Any reactance inserted in the plate circuit will have a different effect from the same reactance inserted at the cathode circuit.

One point has been overlooked however. We do not normally insert reactances in one of the output circuits that we do not put in the other. If one tube uses a 0.1-μf capacitor for coupling, the other one will invariably use the same value. Also the grid-to-ground resistors (or grid-to-bias-point resistors) for the output tubes will invariably be identical too. Stray capacitance between the output tube grid and ground will usually be similar.

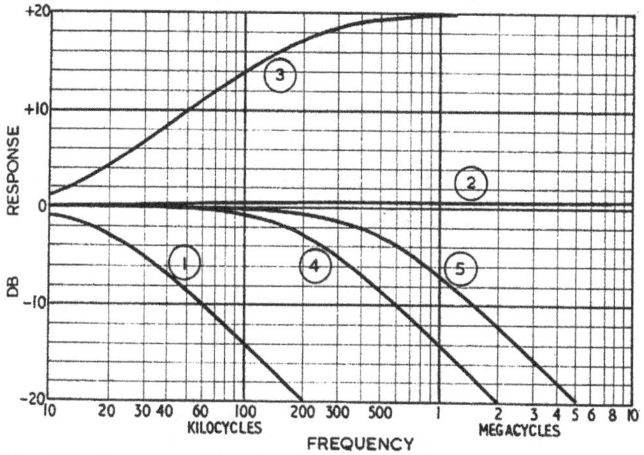

Fig. 406. *Responses showing the effect of shunt capacitance from plate and cathode of Fig. 405: Curves 1 and 2 show response at plate and cathode respectively when capacitance is connected in plate only. Curves 3 and 4 show response at plate and cathode only. Curve 5 shows response when equal capacitance is connected at both plate and cathode, both places giving the same response.*

If any of these components acted *only in one circuit and not in the other* then an unequal effect on frequency response would occur. But, because both happen simultaneously, balance is maintained. Because the plate and cathode circuit of the split-load inverter each contain an identical network of impedances *at all frequencies* the voltage across the two halves of the load is equally divided.

Consider the effect of capacitance across each section separately, not only on the output at that point, but on the other half too. A capacitance at the plate causes a rolloff at the plate and a very slight boost at the cathode, due to the change in effective gain as a partial cathode follower. A capacitance at the cathode causes a much more remote rolloff at the cathode but a very considerable boost at the plate, due to the effective removal of cathode circuit degeneration by a bypass capacitor. This is illustrated in Fig. 406.

It is evident that the response at the plate is subject to much greater fluctuation than that at the cathode. But if the impedances in both circuits are equal, the response at both points will be flat.

This statement does not apply to distortion that occurs due to grid current flowing in the following stage. Obviously the cathode is better equipped to supply grid current than the plate, due to its lower effective source impedance. In this case the tube currents do not flow simultaneously, but at opposite peaks, hence the balancing effect noted with regard to frequency-response parameters does not apply to distortion cancellation.

Distortion in the split-load inverter itself, can be deduced from the distortion produced by this tube with the operating load, divided by the negative feedback factor between grid and cathode. However, the fact that a large swing is required to drive the split-

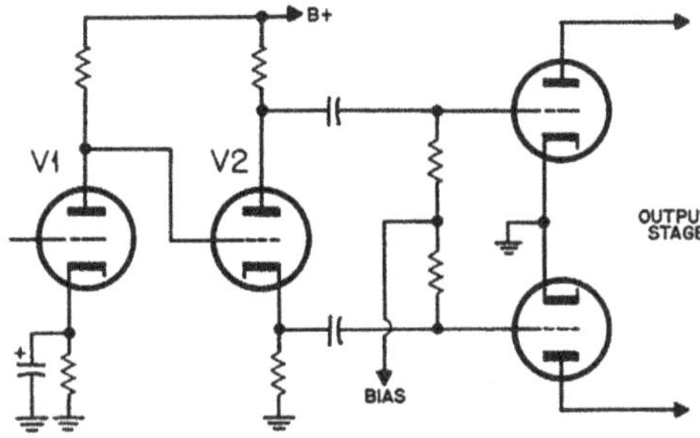

Fig. 407. *Split-load inverter with a direct-coupled preceding stage.*

load inverter may mean that the previous stage needs to be operated under conditions that involve higher distortion than would be necessary with an inverter in which there is some gain.

Where the split-load phase inverter immediately precedes the output tubes, it does not provide convenient points for local-loop feedback from the output tube plates. However, if it is used ahead of a separate push-pull driver stage, then local-loop feedback can be used between the output and driver stages independently of the split-load inverter.

Sometimes the split-load phase inverter is seen directly coupled to the plate of the preceding stage. This is possible because its grid is a few volts negative of the dc operating potential of its cathode. For some circuits this is not too low a plate voltage for the preceding stage (Fig. 407).

In this circuit, however, the preceding stage must be considered as part of the phase-inverter circuit because the plate voltage of the preceding stage V1 directly controls the grid voltage of the phase inverter V2 and this must not only get the correct voltage swing to drive the output tube but must have the correct bias so that this swing is available without undue distortion. For this reason it is usual to control the voltage of the preceding stage by using a rather larger cathode bias resistor than usual with a correspondingly large value of plate resistor. This tends to swamp fluctuations in the tube parameters and make the tube a constant-current device from the dc viewpoint. The tube gain, if necessary, can be restored by bypassing the rather large cathode resistor by a suitable bypass capacitor.

Paraphase

The next variety of phase inverter to receive consideration is that called paraphase. This uses an additional stage to provide phase reversal. Most of the remaining types are really variations of the original paraphase. In its simplest form the method consists of using one stage to amplify the signal for the grid of one output tube and then attenuating into a further stage that provides the signal for the opposite output grid. By arranging that the attenuation between the stages is equal to the gain of the second tube, both grids will receive equal voltages but in opposite phase due to the reversal effect in the tube (Fig. 408).

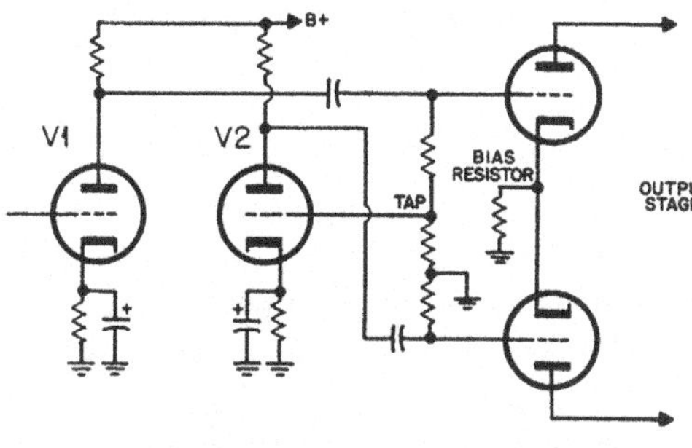

Fig. 408. *Simple "paraphase" circuit.*

The simple paraphase circuit has an advantage over the split load in that each output tube grid receives the swing from a whole plate load. This means that, using the same tube type, the available swing is about twice that of the split-load phase inverter for a given amount of distortion.

In addition, the second tube will provide some distortion cancellation. The first tube, of course, will provide its quota of second-harmonic distortion in the drive to the first output grid, due to the curvature of the tube characteristic. Assuming that the two tubes used in the paraphase circuit are identical, this curvature will be attenuated and then amplified by a tube in opposite phase. The identical tubes will produce equal components of second harmonic but in opposite phase. The second harmonic should cancel in the second paraphase tube so that the second output tube receives zero second harmonic in its drive. Ignoring for the moment the distortion effects of the output tubes themselves, when the signal is recombined in the plate circuit the resultant distortion will be approximately half that in the drive to the first tube grid (Fig. 409).

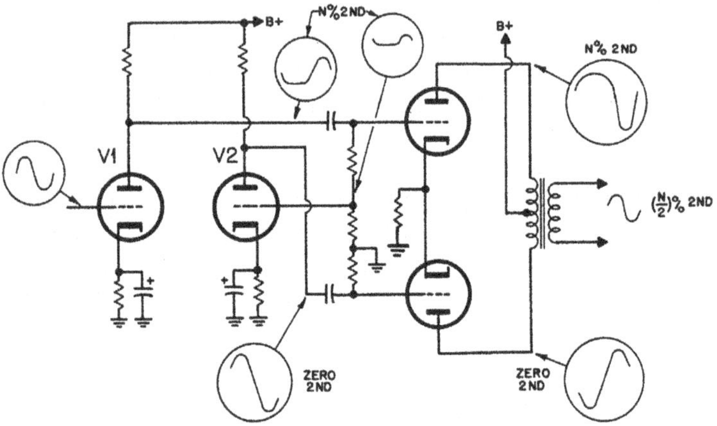

Fig. 409. *Effect of a paraphase circuit on the distortion produced in the tubes.*

The big disadvantage of the simple paraphase circuit is that the balance is entirely dependent upon the tube parameters. If the gain of the second tube changes from the value used in determining the attenuation, then the drive for the output tubes will not be the same.

A further disadvantage occurs in the balance with regard to frequency response. Since there is an extra stage of coupling for the second output tube, this means that it must suffer additional phase advance and loss at the low frequencies and also additional phase delay and loss at the high frequencies. The drive for the two output tube grids is not identical over the whole frequency range nor in exactly opposite phase even though it may be so at a mid-band frequency where these reactances have no effect.

This simple paraphase circuit is of course quite inconvenient for the application of feedback because a local-loop feedback cannot conveniently be injected into either tube without upsetting the balance for the other.

As regards the stability criteria of the complete loop, the extra reactance for the second output tube both at the low and high ends means the circuit constitutes an additional liability in overall feedback design.

Output-coupled paraphase

An interesting variant of the simple paraphase circuit is one in which the output stage is virtually its own phase inverter (Fig. 410). Here the drive for the second output stage (V2) grid is obtained from a tap in the plate circuit of the other output tube (V1). This can be obtained, either from a suitable tap on the primary winding of the output transformer or by a resistance potentiometer across that half of the winding.

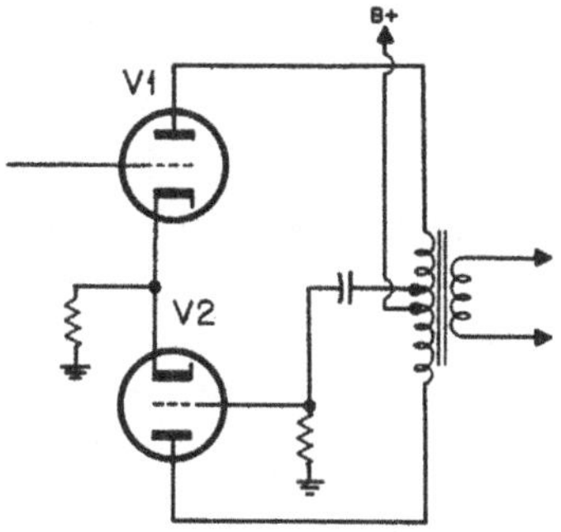

Fig. 410. *Simple output-coupled paraphase inverter.*

There is no particular advantage, performance-wise, to either method. The tap on the transformer saves the necessity for two additional resistors which may need hand-picking for close tolerance values.

Most of the disadvantages inherent in the paraphase circuit apply to this one. From the viewpoint of overall design it might be considered as a single-ended output stage because the push-pull action is derived within the stage itself. Because of this, the additional stage necessary with the normal paraphase arrangement is avoided, hence the additional parameters involved are to some extent not needed. In other words there may be one less stage in the overall loop.

On the other hand the stability of balance against tube parameters is seriously aggravated here because the gain of each tube is dependent upon its receiving the correct plate load, especially where the tubes are pentodes. Practical speaker loads do not provide the ideal load resistance for the tubes. If this load value is higher than nominal, this means that the drive provided for the second tube grid will be larger than that for the first. The second tube will provide a greater proportion of the total output power available than the first tube. The second tube will accordingly run into distortion before the first one and the output will be cut down considerably from the nominal push-pull normally available from the two tubes used. Of course, lightly loading any pair of tubes reduces the available output due to the fact that the optimum load is not used. But the effect just discussed is over and above the loss of power caused by normal mismatching and is due to the fact that the tubes are operating with unequal input swing and with unequal effective plate loading (Fig. 411).

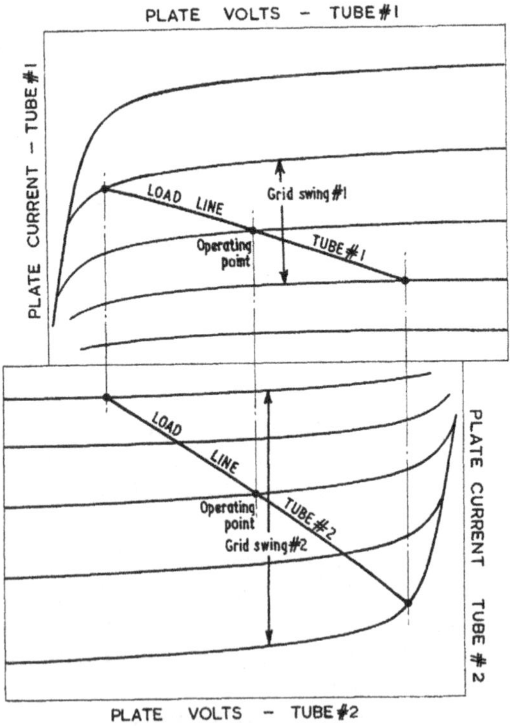

Fig. 411. *How loading affects the individual tubes differently with the circuit of Fig. 410.*

Another disadvantage of this circuit is that the waveform fed to the second tube is the output from the first tube. Closely coupled with this is the output from itself, phase-reversed by the transformer. This will minimize distortion from the second tube but cannot do so from the first. Due to "load-switching"–during different portions of the cycle each tube supplies different portions of the total power (even when correctly matched)—the first tube will introduce more than the usual amount of distortion anyway.

When these two effects combine, as they naturally do when the load value deviates from its theoretical value, this stage produces quite a comparatively high quota of distortion, even at relatively low levels.

Floating paraphase

Two more types of phase inverter can be regarded as modifications of the paraphase type. The first is called the floating paraphase, of which one variant is shown in Fig. 412.

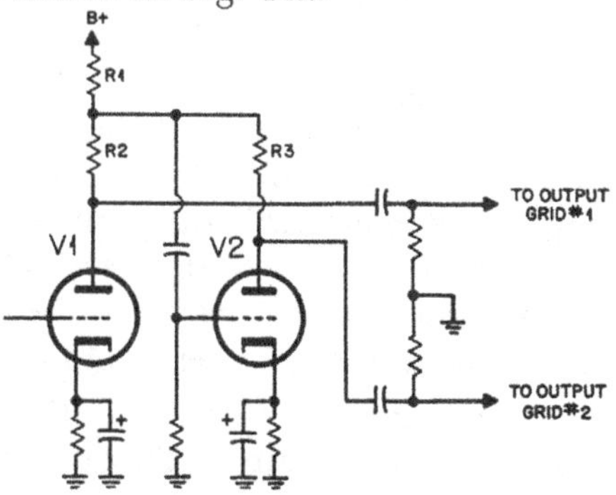

Fig. 412. *Variation of the floating-paraphase inverter circuit.*

Its design hinges on the choice of values for resistors R1, R2 and R3. The drive for the grid of V2, is taken from the junction of these three resistors. If R2 and R3 are equal and the signal voltages at the plates, for delivery to the output tube grids, are assumed equal and opposite in phase, then the current swing through R2 and R3 must also be equal and opposite in phase. This means R1 has no current swing so the junction of the three resistors cannot swing at all. For this reason the grid of the second tube receives no drive. Hence the assumption that the output voltage of the second tube is equal to that of the first tube V1, is impossible.

The current swing of the first tube *must* be slightly larger than that of the second, so the voltage produced by the *difference* in current swing in R1 will give the correct grid drive voltage for the second tube.

To illustrate the advantage of this method of operation, let's put in some figures. Suppose that the current swing of the first tube is 10% greater than that of the second one—that is, the relative swings are in the ratio of 11 to 10. The current swing in resistor R1 will be 1/10th of the swing in the second tube or 1/11th of that in the first tube. By making the resistors in opposite relationship—that is, R3 is 10% greater than R2, the signal swing at the plates of the tubes will be identical again.

Now assume that the transconductance of the second tube drops by 10%. This would mean that, instead of producing 10/11ths of the current swing provided by the first tube, it would produce only 9/11ths and therefore the grid swing existing at the junction of the three resistors would be twice the value it was before. This is impossible because then, even with the 10% reduced transconductance, the output would be 1.8 times its previous value from this tube—or a greater current swing than the first tube.

Of course the circuit readjusts itself by the "floating" effect so that the available grid swing rises by 10% to offset the drop in transconductance. This means that the current swing ratio, instead of being 11 to 10, will increase to 10 to 9 and correspondingly, since the resistors are designed for a ratio of 11 to 10, the output at the plates will differ by about 1% for a 10% drop in the transconductance of the second tube. Thus, the floating paraphase considerably reduces the dependence of the circuit upon the operating parameters of the tubes.

Now we come to the disadvantages. Assuming that the tubes provide a working gain of 20 each, the current passing through R1 is only 1/10th of the current swing in the second tube or 1/11th of that in the first tube. To produce 1/20th of each plate voltage will mean that R1 must be half the value of R3. Since the R1 carries the combined plate currents of both tubes, there will be as much voltage drop across R1 as across R2 or R3.

This circuit results in a serious lowering of the effective B-plus available for the tubes. It is not quite as bad as the split-load arrangement, because the junction does not receive half of the output signal voltage—only about 1/10th. So R2 and R3 do receive almost the entire output of the tubes, but only at the operating B+ voltage present at the junction of the three resistors. This way each output is slightly more than that obtained from the split-load arrangement, but not very much.

The exact relationship will depend on the tube type used for comparison. Take the 12AU7 with a supply of 300 volts: for the split load inverter, a load of 50,000 ohms and a bias of −10 volts gives an operating point of 180 volts, and a total swing from 50 volts to 260 volts gives 105 volts peak-to-peak at both plate and cathode; the paraphase will supply the full 210 volts at each plate;

but the floating paraphase will lose half the operating drop in the common resistor, so we use an effective plate supply of 240 volts with a load resistor of 25,000 ohms; biasing to –7.5 volts, the available output swing per plate is from 60 to 210 volts, or 150 volts peak-to-peak.

There is an advantage over the split-load arrangement in that the tubes have an effective gain. The gain will actually be quite similar to the combined arrangement of two tubes, the second of which is a split-load inverter (Fig. 407).

With this circuit the balance in frequency response is a little better than the standard paraphase circuit because there is effectively a degree of negative feedback operating over the second stage. This happens because the voltage applied to its grid is derived from the current swing of the two tubes combined in R1, hence part of the voltage applied to the grid of this stage is due to the current in its plate circuit. This means that the additional coupling parameters, due to stray capacitance from the junction of R1, R2 and R3 for the high end and the coupling capacitor between this junction and the grid for the low end, have their effective frequency band extended by the apparent amount of negative feedback.

This extends the range of frequency over which the phase inversion maintains effective balance. It also improves the problem of stability criteria over a complete feedback loop compared to the simple paraphase circuit.

As the tubes use separate bias resistors it is feasible to arrange local-loop feedback from the plates of the output tubes to the cathodes of the floating-paraphase tubes.

This circuit produces distortion cancellation inferior to the direct-paraphase arrangement. Not too obvious, it can be deduced as follows: The voltage for the second tube grid is obtained by the difference between two currents that would contain almost equal second-harmonic percentages, assuming that the voltages at their grids are the same waveforms but antiphase. Although the fundamentals almost cancel in R1, the second-harmonic components would be additive. But feedback via the second tube grid tends to cancel this combined second-harmonic component of current. It cannot reduce the actual second-harmonic component of current swing in the first tube but will set up antiphase current in the second tube to neutralize it. Thus the signal for the output tube grids will contain approximately equal proportions of second harmonic but in opposite phase—the same as the fundamental.

This will not cancel in the output.

Another way to view this is that the second tube of the floating paraphase acts as a feedback *current* inverter for the plate current of the first tube, which has no feedback. Thus the harmonic generated by the first tube gets inverted along with the fundamental.

Another variation of the floating paraphase puts the critical resistors in the grid circuit of the output stage instead of the plate circuit of the paraphase couple (Fig. 413).

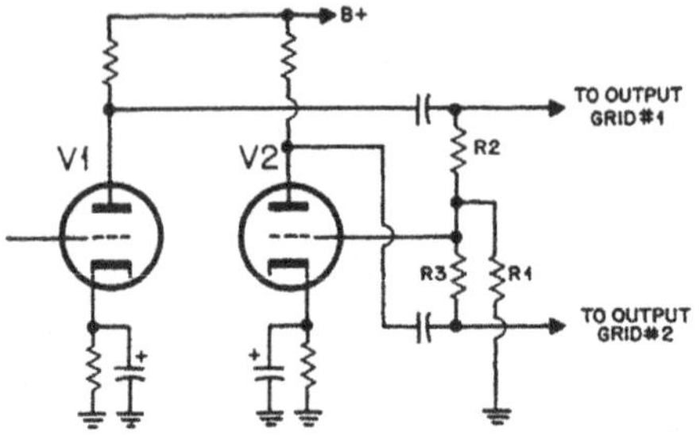

Fig. 413. *Another variation of the floating-paraphase circuit.*

However this circuit does provide a larger swing because the full supply voltage is retained; here R1 has a difference signal in the grid circuit, which carries no dc. There is in practice a very slight loss compared with the straight paraphase because the output tube grids must have not more than a maximum grid circuit resistance. As R1 is common to both circuits, this means that R1, R2 and R3 must be about one-third of the permissible value. For the straight paraphase the full value can be used in each grid. This will shunt the gain of the phase splitter slightly, depending on the tube type and the value of maximum resistance for the output tubes used.

It is possible to eliminate the capacitor between the junction point and the grid of the second tube if ground is the bias return point for the grids of both the second tube and the output tubes. This does not mean that a low-frequency phase shift has been removed. There is still a phase advance in each of the coupling capacitors from the plates to the resistor triad. This means that the feed to the second output tube grid has virtually passed through one more capacitor in the chain than that to the first output tube grid. However, the presence of negative feedback has effectively increased the value of the additional capacitor. This means that the difference in phase at the low-frequency end is reduced by the effective amount of feedback provided by floating paraphase action.

Another variation of the floating-paraphase circuit, shown in Fig. 414, is sometimes called the "seesaw" circuit. This overcomes one deficiency of the previous circuit. If the output tubes run into grid current, they will temporarily make the junction point of R1, R2 and R3 (Fig. 413) negative, due to the change of charge on the coupling capacitors. This may change the operating condition of the second phase-splitter tube and can even run it momentarily to cutoff in a serious case. Whether this effect is as serious as the change of bias on the output tubes which accompanies it and whether it causes crossover distortion due to overbiasing depends on individual circuit parameters. The change of circuit in Fig. 414 does not guarantee freedom from effects of this general nature.

Long-tailed inverter

Another variant that achieves a similar objective to the floating paraphase is given the name long-tailed inverter, the circuit of

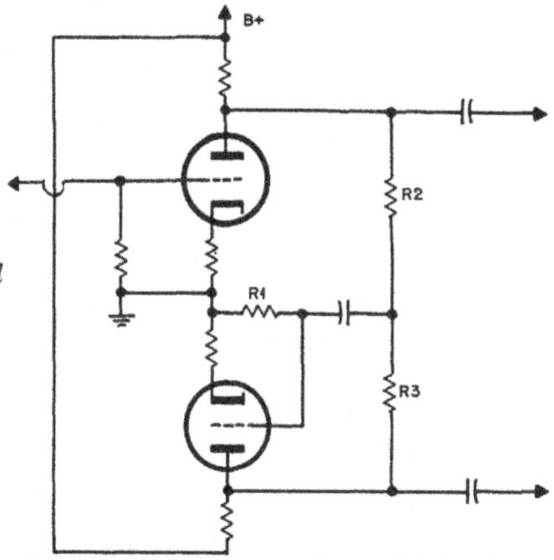

Fig. 414. *See-saw circuit is modified floating paraphase.*

which is shown in Fig. 415. Here the resistor, in which the currents of the two tubes are combined to provide a difference signal for driving the second grid, is in the cathode circuit of the two tubes. This means that the plate resistors return to B+.

However, there is the same reduction in available plate voltage for the same effective reduction in dependency on tube parameters as in the circuit of Fig. 412, because the same voltage drop will occur across R1 whether it is in the plate or the cathode of the combined tube circuit. So the handling capacity is precisely similar whichever arrangement is used.

Where a big swing is needed, the output tubes usually use fixed bias, which means that a negative supply is available. By careful design, this can be used for the long-tail return so that the grids of the phase inverter can operate around ground potential. This saves the loss of B-plus and allows the full paraphase swing to be obtained. The circuit is shown in Fig. 416.

Stability of balance against tube parameters is similar because the derivation of the drive for the second tube grid is still the difference between the currents through the individual tubes. The choice of values is made on the same basis.

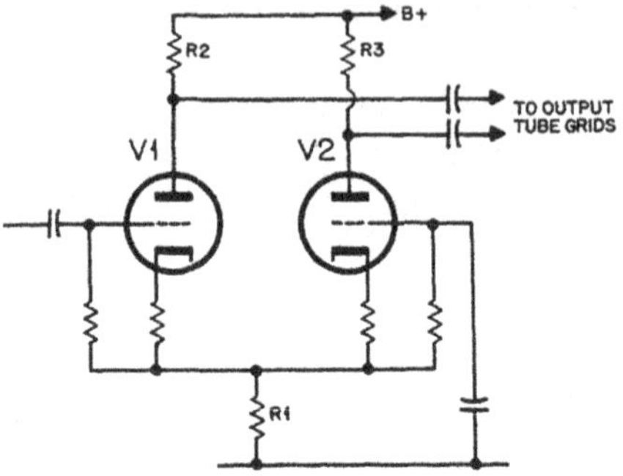

Fig. 415. *"Long-tailed" inverter.*

In this case, however, the grid of the second tube is at ac ground while its cathode fluctuates; in the floating paraphase, the cathode is at ac ground potential while the grid receives the drive voltage. An advantage of this arrangement is that the capacitor from the second grid to ground can be made much larger than is normal

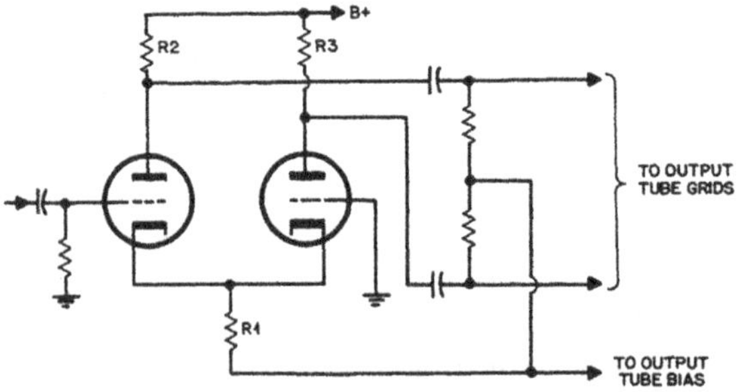

Fig. 416. *Alternative of the long-tailed inverter.*

practice for coupling capacitors. Being at ground potential, it will not introduce additional stray capacitance, thereby affecting the high end. In the circuit of Fig. 416 the second grid is actually connected to ground. This means that its defect in producing an additional low-frequency phase shift can be practically eradicated. The high-frequency phase-shift situation is also considerably improved in the long-tailed inverter because the impedance at the cathode junction of the two tubes is extremely low. Any stray capacitance at this drive point can virtually be ignored as the impedance is probably less than 1,000 ohms.

The distortion of the long-tailed inverter is much improved over the floating paraphase. Not only does this arrangement give about 6-db overall feedback, but second-harmonic components are fed back to *both* grids. The long-tailed inverter virtually provides an automatic division of the input grid voltage into two, the floating point at the cathode being halfway signal-wise between the second tube grid, which is virtually at ground, and the input voltage. Since this reduces comparative gain to half, it is equivalent to 6-db overall feedback.

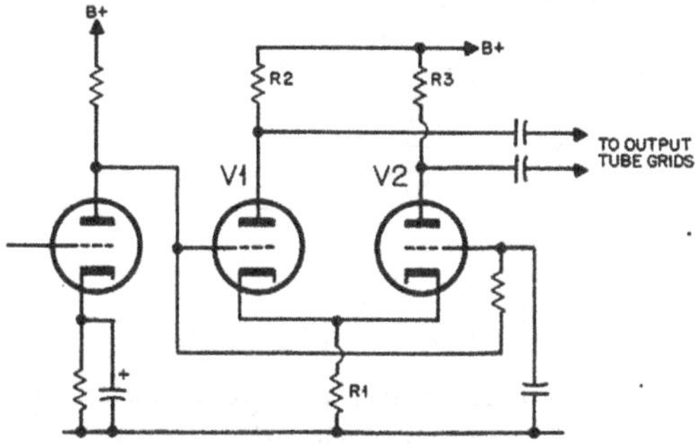

Fig. 417. *Variation of long-tailed inverter improves the stability of the amplifier and saves some components.*

This circuit does not provide any convenient point for applying local feedback from the output stage because the cathode circuit is common to the two tubes.

The stability criteria are somewhat improved for the reasons already mentioned, in respect to the improved balance in frequency response. A further advantage in regard to overall stability criteria of a complete amplifier is obtained by direct coupling from the stage before the long-tailed inverter. This can be achieved with the circuit of Fig. 415 because the grid potential of the long-tailed inverter is above ground owing to the necessity for the "long-tail" resistance. This means that, by a design procedure similar to that used for the twin-stage split-load arrangement, a further coupling capacitor can be eliminated. Also, the bias for the second half of the long-tailed inverter can easily be obtained by using a large-value resistor between the two grids and applying just the capacitor to ground. This considerably simplifies the circuit (Fig. 417).

Cross-coupled inverter

The inversion part of this circuit consists of two tubes to which the signal is fed in opposite phase (Fig. 418). This circuit was originally designed to accept either single-ended or balanced inputs or ones that were neither. It will mix the input at each grid, giving equal out-of-phase components of each at the plates. To see the effect on a signal at one grid, consider the other one to be held at ground potential (i.e., as a single-ended input). Regarding the

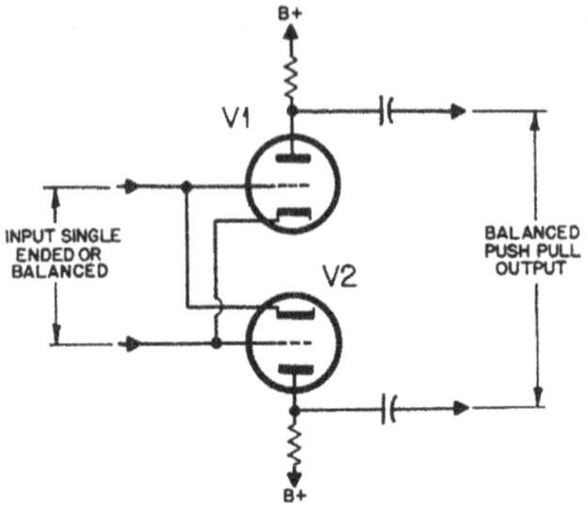

Fig. 418. *Basic circuit of the cross-coupled inverter.*

lower input terminal to be at ground potential, the upper tube acts "grounded cathode," with the signal applied to its grid, while the lower tube acts "grounded grid," with the signal applied to the cathode.

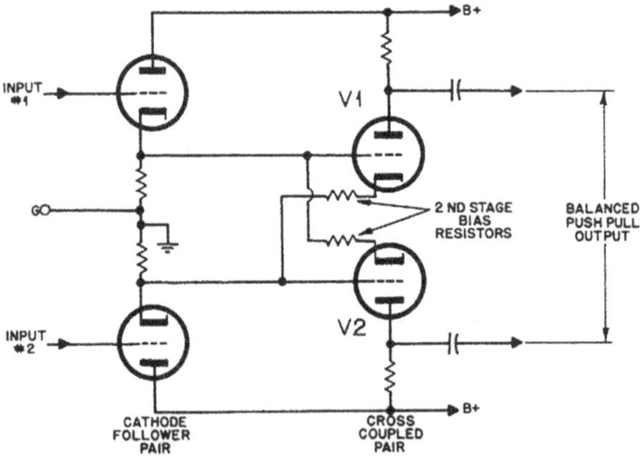

Fig. 419. *Practical circuit for cross-coupled inverter includes cathode-follower pair to drive it.*

Because the load impedance of a cathode, in a grounded-grid circuit, is quite low and that of a grid quite high, this circuit must be fed from a source of low resistance to prevent undue loading and possible distortion. For this reason the cross-coupled inverter is inseparably connected with a push-pull cathode-follower arrangement (Fig. 419).

Basically, this was not developed as an audio circuit but as a suitable input for a vtvm or scope. It can conveniently be used as a balanced or single-ended input—or even for an input which is neither—and always gives a true push-pull deflection signal at the output. It obviates the necessity for changing the amplifier connection in some way to change over from single-ended to push-pull types of input and is a great asset for this application.

It has been applied, by connecting one of the input grids to ground, as an input circuit for audio power amplifiers. It is pos-

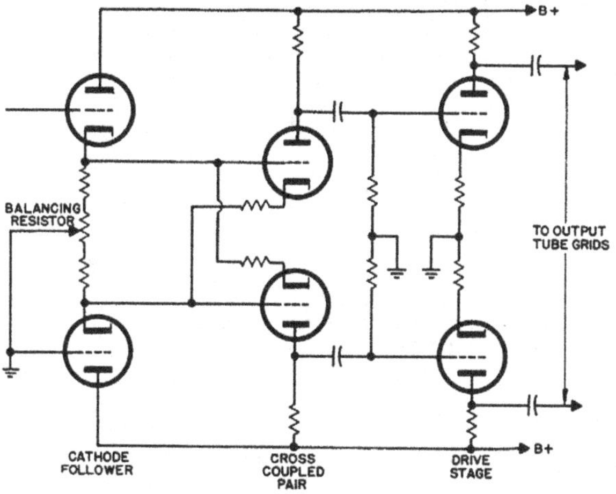

Fig. 420. *This type of inverter works better at the front end, with a driver stage following.*

sibly somewhat uneconomic for the reason that a single-ended audio amplifier does not require this adaptable feature. The input is either balanced or single-ended—usually the latter, and is not likely to change from one to the other at will. The circuit could be used later in the amplifier to provide a direct drive for the output tube grids except that the cathode follower needs to operate at fairly low level, so a subsequent stage of push-pull amplification is usually necessary between the phase inverter and output as shown in Fig. 420.

It is worth a closer look at the balancing method of Fig. 420. The cross-coupled pair phase-splits the signal delivered at either (or both) the first stage cathodes. This means that adjustment of the control can change the balance only by changing the relative gain in the upper and lower channels. It does this by changing the dc component or operating voltages of the cross-coupled pair.

The circuit of Fig. 421 overcomes this objection by using an improved cathode-follower stage, with closely controlled dc equality and making the balancing adjustment in the plate circuit of the phase inverter.

The circuit is inherently stable with normal tube-parameter variations but is susceptible to differences between halves of the 12AX7 or cross-coupled pair. The balancing control takes care of this.

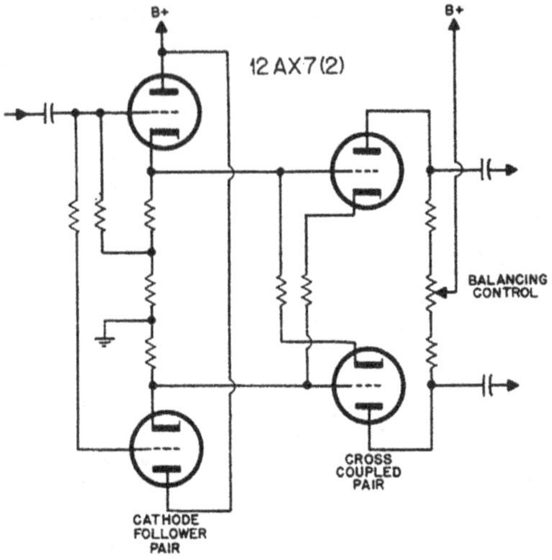

Fig. 421. *Improved cross-coupled inverter.*

Balance over the frequency response is particularly good, because there are no capacitors in the inverter arrangement to produce low-frequency phase shift. The whole arrangement is direct-coupled from the grid of the first input tube to the two phase-inverter plate outputs. From this stage on resistance–capacitance coupling will normally be used and both halves of the push-pull will be identical. This also applies to the parameters affecting high-frequency response.

An advantage of using push-pull amplification over several stages is that, where the grids are at high impedance (those of the second stage of this inverter are not, being connected to the cathode follower), it is possible to use neutralization to overcome the normal high-frequency rolloff parameter (Fig. 422). A capacitance, equal to the plate-to-grid capacitance of the intermediate stage, is connected from the plate to the opposite grid. This means that the normal Miller-effect rolloff is neutralized and the response is extended to an extremely high frequency, such that the parameters of this stage can almost be ignored. This facility is not normally available in audio amplifiers for compensation at the high-frequency end and is rendered possible only by push-pull amplification.

Push-pull amplification throughout also makes it readily possible to apply local-loop feedback from the output stage plates to suitable cathodes of an intermediate stage.

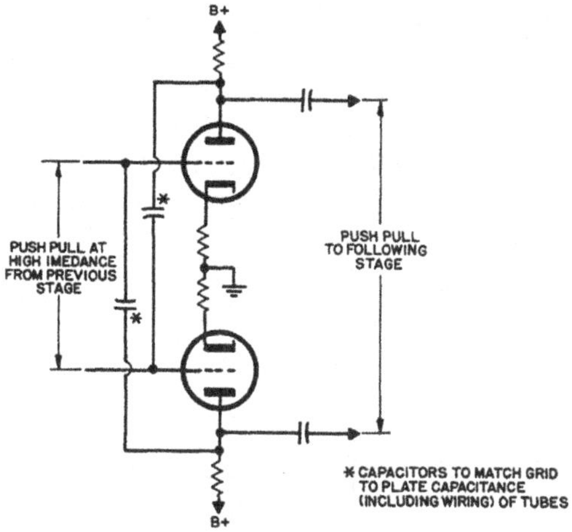

Fig. 422. *How high-frequency neutralizing capacitors can be connected when push-pull triode voltage amplifier stages are used.*

Another feature of this cross-coupled input stage inversion is its suitability for overall loop feedback. This can readily be provided to the grid of the alternate input tube, if desired (Fig. 423). This

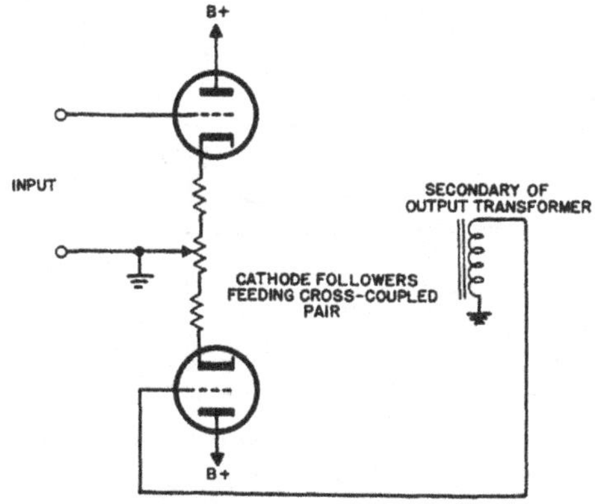

Fig. 423. *Overall feedback connection for the cross-coupled inverter.*

is not often an outstanding advantage because the values, when feeding back to the normal cathode point, are not particularly difficult to attain.

However, with all these advantages to the circuit, an amplifier designed around a push-pull inverter of this type at the front end makes a sort of deluxe basic unit as a toy for audio enthusiasts. From a commercial design standpoint, the arrangement is hardly economic because it necessitates almost twice as many tubes in the power amplifier as well as the associated resistors that go with them.

Extra push-pull drive stages

Due to restrictions in the handling capacity of all the tube types of inverter, any of them may be followed by a push-pull stage of amplification feeding the output tubes themselves, especially where a complete or partial cathode-follower arrangement of output tubes is used.

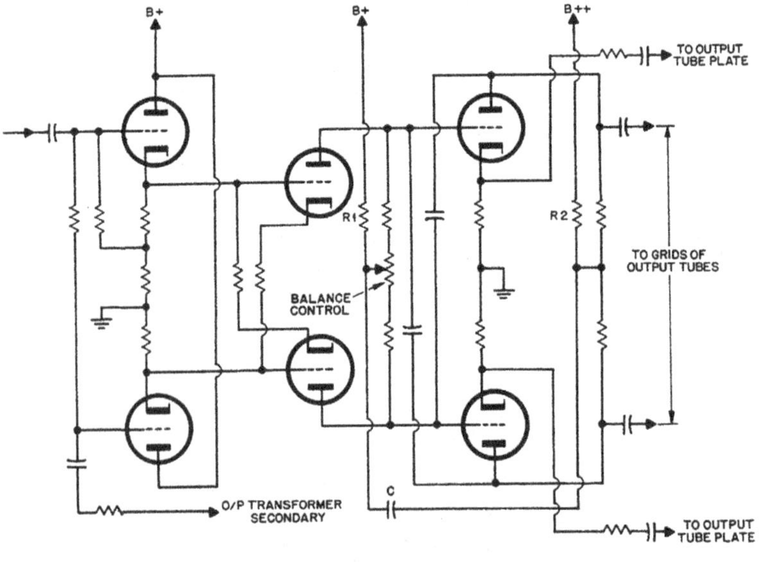

Fig. 424. *Cross-coupled inverter showing how one low-frequency rolloff parameter can be eliminated by direct coupling between inverter and driver stages.*

Sometimes the number of low-frequency parameters can be reduced to two, as shown in the typical cross-coupled arrangement of Fig. 424. The only coupling capacitors are the push-pull ones between the drive stage and output stage grid. These, as a single push-pull low-frequency loss, work in conjunction with the effective primary inductance of the output transformer—the only low-frequency-loss parameters. Similarly at the high-frequency end, the only loss points are at the grids of the output stage and the two elements in the output transformer, primary capacitance and leakage inductance. Capacitance effects at the plates of the drive stage are neutralized by the cross-coupled neutralizing capacitors.

Where there are two consecutive stages of push-pull amplification, it is possible to add a measure of self-balancing. To do this extra resistors R1 and R2 are connected, as shown in Fig. 424. If there is any unbalance in the signal through the amplifier, part of the unbalance will appear across these resistors, but in opposite phase and at a difference in level because of the stage gain. Also any second harmonic generated by the half-channels will be added in these resistors, although the true push-pull signal will neutralize. By coupling the two together, the phase opposition reduces the magnitude of both out-of-balance components and second-harmonic generation, in the manner of normal negative feedback action, using the second stage as a push-push-coupled arrangement while the wanted signal component is in push-pull.

The effective amount of feedback available for this purpose is not as big as may appear at first. The signal from the tie point of R2 is fed back through the capacitor to the grid circuit, where it is attenuated by the individual plate resistors in series and the plate resistance of the previous stage in shunt. From this viewpoint it is better to couple to the tie point of the previous-stage plate resistors rather than to set up a similar junction in the grid circuit (where R-C coupling is used—this cannot be shown in Fig. 424, which uses direct coupling at this point) because the grid resistors are usually larger than the plate resistors hence would provide less effective feedback of what is sometimes called "longitudinal feedback."

The inner feedback loop, shown from each output tube plate to the cathodes of the driver stage, actually has three low-frequency parameters, due to the additional blocking capacitance necessary here. However the values can be adjusted to avoid peaking in this internal loop. The overall feedback loop—unusual as it may be—has only two parameters contained in it apart from any peaking that may be introduced by the inner loop.

From the viewpoint of feedback amplifier design, the two feedback loops (here regarding the push-pull inside one as a single loop) should really be designed complementary. The inner loop has only two high-frequency parameters, the grid-to-grid and plate-to-plate capacitances of the output tubes. The latter includes the primary capacitance of the output transformer. Then the overall loop includes one extra reactance, the leakage inductance of the output transformer.

At the low-frequency end, care needs to be exercised more particularly on the inner loop because this uses three low-frequency reactances. Once this has been rendered adequately stable with a good rolloff margin, the outer loop does not present problems because the effective reactances here will only be two—provided the inner loop has been satisfactorily taken care of.

Positive feedback drive

Now we come to a particular type of drive stage in which positive feedback is used to help get the necessary swing from a relatively small tube at a low plate voltage. This kind of arrangement can be used with the Circlotron or unity-coupled arrangement or

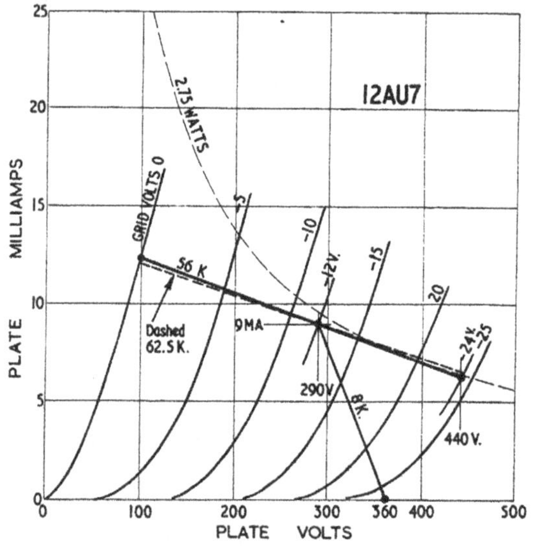

Fig. 425. *Positive feedback increases the drive obtainable from the 12AU7.*

any type of amplifier where there is a high push-pull voltage at the plates of the output tubes and a comparable push-pull drive required at the grids.

Using the Circlotron circuit, let's pursue the possibility of using a 12AU7 with a plate supply voltage of 360. The appropriate characteristics are shown in Fig. 425, a suitable operating condition that will just about give the required swing using an 8,000-ohm resistor and a bias of 12 volts. This gives an operating condition of 290 volts on the plate at 9 ma of plate current. So the bias resistor will need to be about 130 ohms and the operating point close to the maximum dissipation of the 12AU7 as well as its maximum plate working conditions—2.75 watts and 300 volts, respectively.

When the output stage has its correct plate load, the effective positive feedback is 7.2. Using, for convenience, the round figure of 7, this means the effective plate load resistance, through the operating point, will be about 56,000 ohms. Drawing this load line and estimating the swing for an input from zero to −24 volts, we find that the plate swings from 290 to 100 and up to 440 volts. A peak-to-peak swing of 330 volts is necessary, so this arrangement gives a working margin of 10 volts.

If the plate load resistance of the output stage should be larger than its nominal value, the swing required at the grid will be increased correspondingly. This means also that the positive feedback will be increased proportionately. However, from the load line in Fig. 425, increasing the slope does not proportionately increase the length of the load line at this point. Consequently the use of a load higher than the nominal value on the output tubes, will mean that this tube will no longer be capable of providing the necessary swing.

For this reason a better tube, such as the 12BH7, which can utilize a higher plate voltage, is an advantage. The pentode at the output can also use a higher supply voltage but there will be an overall gain, providing a better margin.

But, just to complete the picture, assuming the correct nominal load is used, let's continue the figuring for the 12AU7. The swing produced by each half is 190 volts in one direction and 150 in the other. Thus there is an off-center effect of 20 volts in a peak-to-peak of 340 volts, representing about 6.2% second-harmonic distortion. However this 6.2% will be in opposite phase on the two halves of the push-pull driver stage and, other things being equal, will cancel at the plate circuit of the output stage.

However the fact that it will interact with the distortion of each output tube, which also will possess some nonlinearity of an opposite kind, will complicate matters somewhat and tend to produce higher-order-harmonic components of somewhat lower level.

This assumes that the positive feedback multiplies the actual plate load resistor by the factor of 7 *as a linear multiplier.* The nonlinearity of the output stage will mean that this factor of 7 is not constant so the load resistance applied to the 12AU7 will not

Table 4-1. COMPARATIVE QUALITIES OF PHASE INVERTERS

Type of Phase Inverter	No. of stages	Relative O/P swing per side as fraction of maximum as V/A stage	Relative voltage gain of typical stage	Distortion in O/P as fraction of amount as V/A stage†	Stability of balance with tube-parameter variation	Frequency Response	Balance in Frequency Response	Suitability for use with feedback: Local Loop	Rolloffs Overall* L/F	Rolloffs Overall* H/F
Direct-coupled transformer	1	3	30	smaller	excellent	can be good	can be good	yes	1	3
Parallel-fed transformer	1	3	30	smaller	excellent	can be good	can be good	yes	2	2
Tapped af choke	1	1	10	same	excellent	can be gooa	fairly good	no	2	1½
Split load	1	0.5	0.8	10–20%	excellent	excellent	excellent	no	1	1
Split load preceded by direct-coupled stage	2	0.5	10	same	excellent	excellent	excellent	no	1	2
Simple paraphase	2	1	12	50%	poor	good	good	no	1½	1
Output-coupled paraphase	1	1	12	same	very poor	good	good	no	1	1
Floating paraphase	2	0.7	10	same	good	good	very good	yes	1½	1½
Long-tailed inverter	2	0.7	5	10–20%	very good	very good	very good	no	1	1
Cross-coupled inverter	4	less than 1	12	almost zero 2nd	can be good	excellent	excellent	yes	1	1
Split load with following push-pull stage	3	1	6	10–20%	good	excellent	excellent	yes	2	2
Push-pull stage with positive feedback from O/P	3	10	45	10–20%	good	good	excellent	yes	2	2

* Number given indicates number of rolloffs contributed by the phase-splitter part of the circuit only. In this column, ½ indicates a rolloff in one side of the phase inverter but not in the other.

† Distortion figure given only indicates the contribution due to individual curvature of phase-splitter tubes, assuming identical types, where more than one is used.

O/P = output; V/A = voltage amplifier; L/F = low frequency; H/F = high frequency.

be a constant resistor of 56,000 ohms. However, the fact that the load line is relatively horizontal will mean that the nonlinearity of its value will not add a proportionate nonlinearity to the output voltage.

The positive feedback will, however, multiply the residual distortion achieved by the output stage. In this particular example, the use of positive feedback between output and driver just offsets the negative feedback produced by the splitting of the output load. This means that we shall land back at a distortion figure of approximately 2% before any overall feedback is applied, as in the case of the tube with straight pentode operation.

While the position on the matter of distortion will be restored by this positive feedback to that without the negative feedback due to the cathode circuit, the same effect does not apply to the plate resistance presented at the output. Fig. 425 shows that changing the slope of the dynamic load line on the 12AU7 (such as occurs by removing the output load) will not materially affect the swing it delivers for a given grid voltage input. The plate resistance of the 12AU7 at the operating condition chosen is about 8,000 ohms and the load resistance is 56,000. Removing the output load, with a source resistance of 0.13 the load resistance will increase the output swing so the positive feedback factor changes from 7 to 7.8, which will raise the 12AU7 load line from 56,000 to 62,500. This will increase the swing delivered by the ratio $\frac{62.5}{62.5 + 8} \times \frac{56 + 8}{56}$, or about 1.2%. This change in drive will thus effectively increase the source resistance of 0.13 by 1.2% to about 0.132. So the effective source resistance of the output of a Circlotron or unity-coupled arrangement will still be less than 0.15 times the load resistance.

The distortion will be in the region of 2 to 2.5% so some overall feedback can be used for distortion reduction.

At this point the stability criteria of an arrangement such as this is quite simple to determine. The positive feedback utilizes only the coupling between the plate and grid of the output stage. The coupling from plate of the output to plate of the driver is directly through the plate load resistor. This being the case, there is only one rolloff parameter affecting both low- and high-frequency stability criteria. The presence of positive feedback will bring these rolloffs proportionately nearer to the passband while the negative feedback due to the cathode section of the load on the output tubes will correspondingly widen it. The net result will leave the working parameters approximately in the same position as with a pentode tube directly driven without these positive and negative feedback loops.

While the Circlotron amplifier (Fig. 426) has been used to discuss the merits of this kind of back coupling with a drive stage, the same principle can be calculated for other circuits.

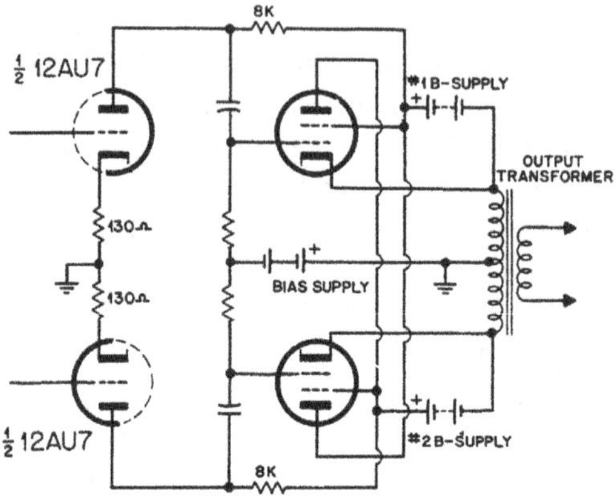

Fig. 426. *Complete circuit of Circlotron output.*

It should be mentioned that, as the tubes and operating conditions are not those of the amplifiers made by Electro-Voice (proprietory name, Circlotron), the figuring given in this book does not represent the actual performance of Electro-Voice amplifiers. Different figures have been used deliberately so as to provide a more definite comparison between the operation of the various circuit types considered. The Electro-Voice amplifiers use different tube types and operating conditions, optimized for the overall application.

Table 4-1 lists the comparative qualities of various types of phase inverters.

chapter 5

input stages

THE problems peculiar to input stages arise because the signals to be amplified are so very small. This is aggravated by the dynamic range to be handled. Suppose a microphone or pickup delivers 2 millivolts of signal at the grid of the input stage. The rated input is probably somewhere near the maximum signal that it delivers: if the system handles a dynamic range of 60 db (which means the loudest signal is 60 db higher in level than the quietest), then the quietest signal—which should still be heard above background noises—will be in the region of 2 microvolts.

This is a very small value and it is quite easy to pick up various components of spurious signals, which come under the general classification of noise. There are two main regions to be concerned with in the input stages. First, the input circuit itself, before any amplification occurs, and second, the way in which we amplify until the signal is big enough not to be bothered with noise problems.

In the academic sense noise is any unwanted signal. It can be classified into three groups. The first is random noise, which may be described as a random sequence of charge movements (due to the random behavior of electrons) which produces sound characterized by hiss. Actually, random noise is distributed throughout the entire spectrum but, because the ear is more sensitive to frequencies above 2,000 cycles, the sound we associate with this kind of noise is hiss. If the noise is passed through a filter that emphasizes the lower frequencies at a slope of 6 db per octave or more, then the hiss gives place to more of a roar. In fact, random noise can be colored by passing it through an electronic filter to give almost any effect we may choose. Sending it through a narrow tuned circuit of, say, 500 cycles gives a sound very similar to a person whistling that particular tone.

When the word "noise" is used without other qualification, more often than not random noise is the form intended. Sometimes it is used in its more academic sense to include all forms.

The second group of noise or spurious signals is hum or other forms of stray pickup. Cross-talk from an undesired program source would come under the same general grouping, but normally in audio work the problem is hum from various stray supply fields.

The third is microphony. This is most commonly associated with the input tube and causes it to emit a "pong" sound in the output when tapped with a pencil or the finger.

However, there are other forms of microphony such as that which occurs sometimes with an input cable as, for example, with the lavalier type mike worn on the chest of the speaker. The cable runs over the speaker's clothing and, if special attention is not given to obtaining a grade of cable free from this effect, any movement of the speaker is accompanied by a kind of rubbing noise picked up by the cable itself. While some of this noise may be transmitted mechanically or acoustically to the microphone as a transducer, some of it is picked up in the wire electrically.

Input circuit thermal noise

The first thing to appreciate about noise of the random variety is that there is a basic minimum due to thermal or temperature effect. This is a property of all resistance materials. It can be expressed as a noise voltage, which means the rms output voltage produced by any resistance is given by the expression

$$e_n = \sqrt{4KTfR}$$

where K is Boltzmann's constant (Joules per degree Kelvin), T the temperature in degrees Kelvin, f the bandwidth in cycles and R the value of the resistance in ohms.

For an everyday formula, since we will not be bothering with a wide range of temperatures, the following approximate formula can be taken, which avoids the repeated use of temperature and Boltzmann's constant,

$$e_n = \sqrt{fR} \times 1.26 \times 10^{-10} \text{ volts}$$

From this we can calculate and tabulate the effective noise voltage generated by different resistance values over various bandwidths. Here is a second reason why random noise tends to sound more like hiss: most of the energy is concentrated in the higher frequencies. For example, there is as much noise energy between 5 and 10 kc, which is one octave musically, as there is between 0 and 5 kc—all the frequencies below this octave.

This basic thermal noise is a *minimum* in any circuit. Sometimes there will be a greater noise than this—current flowing through a resistor, for example, may produce additional agitation or fluctation in the current path. For this reason it is important to exercise care in selection of resistors for input circuits. They should be of particular low-noise or high-stability construction. The regular composition resistors are more noisy than the figures indicated in Table 5-1.

Table 5-1. R. M. S. NOISE VOLTAGES (IN MICROVOLTS) AT NORMAL TEMPERATURE (THERMAL COMPONENT ONLY)

Resistance (ohms)	Bandwidth in cycles 5,000	7,000	10,000	14,000	20,000
1,000	0.29	0.35	0.41	0.49	0.58
2,000	0.41	0.49	0.58	0.69	0.82
3,000	0.5	0.6	0.71	0.84	1.0
5,000	0.65	0.77	0.92	1.1	1.3
7,000	0.77	0.92	1.1	1.3	1.55
10,000	0.92	1.1	1.3	1.55	1.8
20,000	1.3	1.55	1.8	2.15	2.6
30,000	1.58	1.88	2.2	2.62	3.1
50,000	2.05	2.44	2.9	3.45	4.1
70,000	2.44	2.9	3.45	4.1	4.9
100,000	2.9	3.45	4.1	4.9	5.8
200,000	4.1	4.9	5.8	6.9	8.2
300,000	5.0	6.0	7.1	8.4	10.0
500,000	6.5	7.7	9.2	11.0	13.0
700,000	7.7	9.2	11.0	13.0	15.5
1,000,000	9.2	11.0	13.0	15.5	18.0

It does not matter what form the resistance takes, the noise given by the thermal noise formula will still occur. It may be due to the resistance of a voice coil or the windings of a transformer or even the transformer core losses referred as a shunt resistance. All these sources produce thermal noise.

A stepup ratio in a transformer increases the noise voltage by the same ratio that it transforms the signal voltage. This would naturally be expected, but the result is the same as if a higher resistance source were used. Suppose a 50-ohm voice coil is transformed by means of a 40-to-1 stepup transformer and the bandwidth is 10 kc. The 50-ohm voice coil generates .092 μv of noise. A 40-to-1 stepup from this raises the level to 3.7 μv. An 80,000-ohm resistor, which is the effective resistance presented on the secondary of a transformer giving 40-to-1 stepup from 50 ohms, also shows a noise voltage of 3.7 μv. So the noise presented on the secondary is the same as if an 80,000-ohm voice coil had been used in place of the actual 50-ohm unit.

The only increase in noise that the transformer can produce is due to any resistance it may add of itself. For example, if the transformer is 80% efficient, measured by power-transfer efficiency (a typical value for a small input transformer), then 20% of the noise energy produced on its secondary side will be from the transformer and 80% from the actual input source. This means there will be a 25% increase in noise energy, or about 12% increase in noise voltage, relative to the signal voltage, due to the use of the transformer.

So the essential features of the input circuit, from the viewpoint of resistance noise, are

1). Any necessary resistors, such as the feed resistor for a carbon or a capacitor microphone, should be of a low noise type.

2). The attenuation effect of resistors should be a minimum; the efficiency of a transformer should likewise be a maximum.

This means that gain control, where the input is at a low signal level, should be effected after some amplification has raised the signal to an adequate margin above the noise level. This does not apply to inputs where the signal level is well above noise level in the first place.

Hum

Hum can be picked up by one of two kinds of field inducing it: either a magnetic field causes electromagnetic induction in the input circuit or a static field causes electric induction. Technically, both forms produce a voltage but it is simpler to visualize the behavior by thinking of electromagnetic induction as producing a current and electric induction a voltage.

Electromagnetic induction is more readily noticed in low-impedance circuits. It requires a completed loop in which to induce a voltage. If the connecting lead for a 50-ohm microphone picks up a magnetic field from the supply, the field will induce a few microvolts (or millivolts, according to intensity and the size of the pickup loop in the input circuit). When this input circuit is stepped up by an input transformer, the voltage induced on the primary side will be multiplied by the ratio of the transformer (Fig. 501).

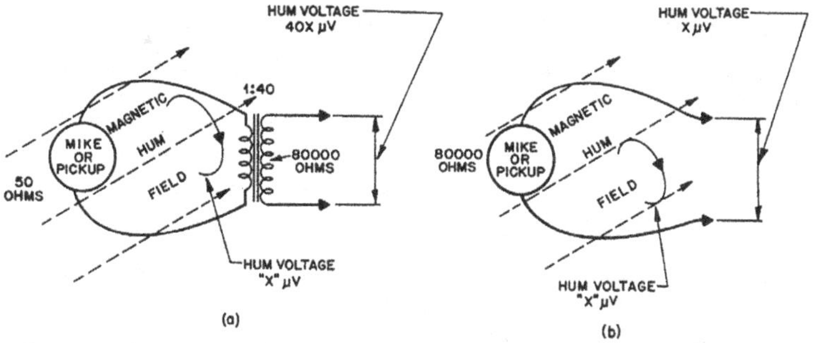

Fig. 501. *Electromagnetic pickup of a magnetic hum field is more serious in low-impedance circuits. The same voltage is induced in the same wiring loop, whether the impedance is 50 ohms (A) or 80,000 ohms (B). But the input transformer steps up the hum voltage as well as the signal, at A.*

In a high impedance circuit, say 80,000 ohms, the same magnetic field would still induce only the same voltage as in the 50-ohm circuit. So use of a 50-ohm circuit followed by a stepup transformer results in induction of 40 times the voltage *as far as the amplifier is concerned.*

Electric induction induces a charge and, if the impedance of the circuit is comparatively low, the charge will be conducted away almost as rapidly as it is induced. Consequently, electric induction is most effective in causing trouble in high-impedance circuits.

Fig. 502. *Electric induction, normally only troublesome in high-impedance circuits, sometimes causes trouble in low-impedance circuits.*

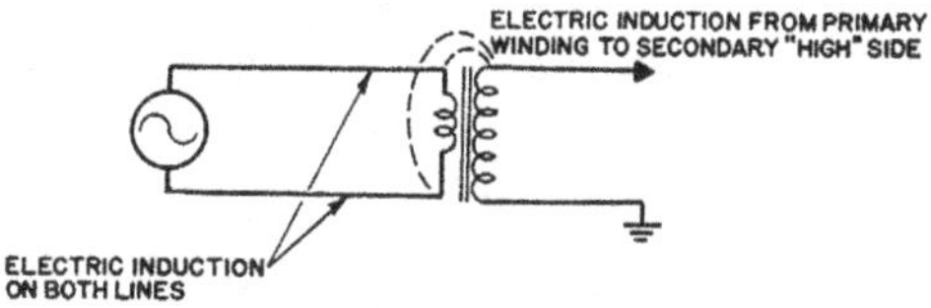

There is, however, a particular way in which it can cause trouble in low-impedance circuits, too (Fig. 502). If the low-impedance circuit is not connected to ground, then the electric field will induce a like voltage in the entire circuit—both sides of the low-impedance line. This will not be transferred to the secondary of the transformer by normal transformer action but, if there is some capacitance between the primary and the high or hot side of the secondary, electric transfer may occur from the low-impedance circuit to the high side of the high-impedance circuit, without true transformer action.

In this simple case the trouble can be cured very simply by grounding the low-impedance circuit. Technically, it should be grounded to a center point to obtain perfect neutralization of any pickup of this nature. But, in practice, grounding either side will be almost equally effective. If the electric induction is very intense, it may be of advantage to ground the center point either by using a tapping on the transformer primary or by putting a potentiometer across the primary and adjusting for a null (Fig. 503).

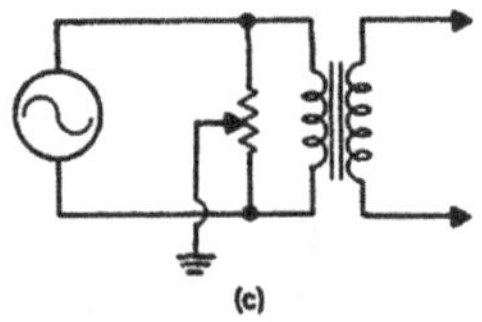

Fig. 503. *Cures for the trouble of Fig. 502. The low-impedance winding should be grounded.* a—*ideal arrangement, connecting through a center tap.* b—*where there is no center tap, connecting one side to ground is often as good.* c—*an artificial center tap.*

However, sometimes the low-impedance circuit has a ground. Maybe there is a connection between the program source and the amplifier, and the former is already grounded. There may be an appreciable hum voltage between the ground of a program source and that at the amplifier. Where grounds are separated by physical distance this is almost a certainty. Under these circumstances the electric field will be set up between the primary and secondary of the input transformer and there will be a considerable electric pickup of hum in the amplifier.

This requires special attention to the input circuit. One means of overcoming the difficulty is to use a line transformer to isolate

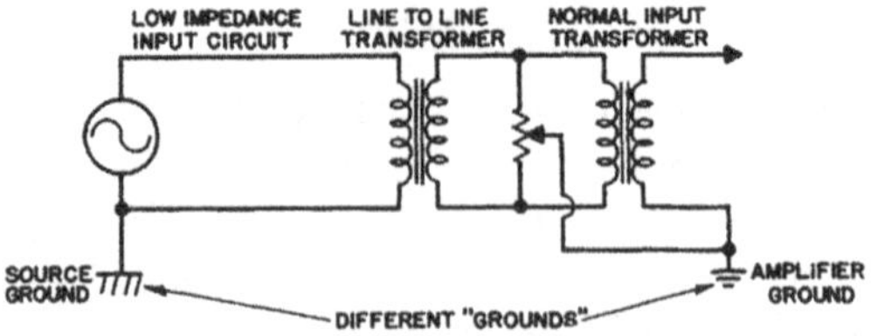

Fig. 504. *Correcting for electric induction caused by unavoidable separate grounds.*

the low-impedance winding from the input transformer itself. Then the intermediate circuit can be grounded back to the amplifier ground (Fig. 504). In this case it is advisable to place a potentiometer across the intermediate circuit and adjust for a null in the electric pickup transferred.

Choice of impedance

Where fairly long connections of, say, more than 200 feet, are used between a microphone and an amplifier, it may be desirable to use a line-impedance connection because a low impedance, such as 50 ohms, will be susceptible to electromagnetic induction while a high impedance will be susceptible to electric induction or high-frequency loss. The middle impedance, in the region of 250 to 600 ohms, is an ideal compromise for such long-distance circuits.

For relatively short distances, up to 100 or 200 feet, the use of this special line impedance is hardly necessary. When using a low impedance, say not higher than 50 ohms, the circuit can be run either in concentric cable—that is, a single shielded wire—or in a fairly tightly twisted pair. This will obviate the loop effect that causes magnetic induction. Either of these are better than twin shielded wire where the twin is flat and not twisted. The shielding on a twin shielded has little effect on the *magnetic* field so the parallel pair still makes an overall loop effect.

For long-distance line-impedance connections, twin shielded should be used and the circuit should be balanced; that is, center-tapped to ground, at one end preferably. The twin shielded should use a twisted arrangement inside the shielding.

High-impedance input connections should always be in single shielded, preferably in some kind of low-loss or low-capacitance-per-foot type of cable. The shunt capacitance introduced by the connecting cable results in loss of high frequencies in a high impedance circuit. There is one exception to this—when a capacitor or piezo-electric type of transducer is connected.

In an input circuit that uses a transformer, this invariably will need shielding, particularly against magnetic hum fields. Any magnetic hum field that reaches the core of the transformer will induce voltages in the transformer secondary, just as current in the primary will.

Two kinds of shielding are effective against a magnetic field. The magnetic shield, which provides an alternative path around the transformer for the magnetic field to follow, is shown in Fig. 505. The essential feature is that the shield shall fit well to provide a uniform path so as not to introduce a further leakage field inside the shield. The shield must be of high-permeability and high-resistivity material so that fluctuations in the magnetic field are not delayed by the shielding material. For this reason the shield is most effective at low frequencies.

Some hum fields, particularly those due to full or partial sat-

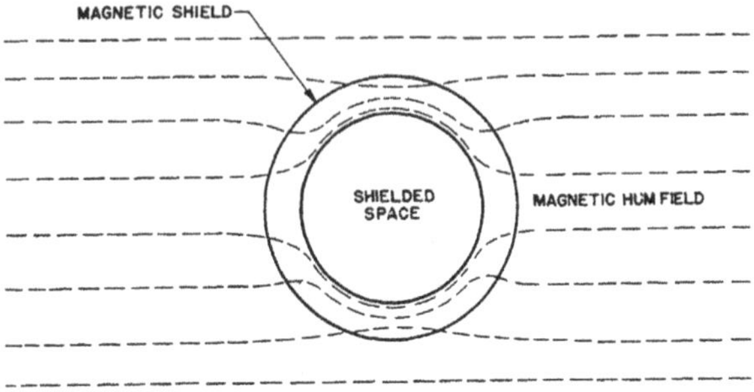

Fig. 505. *A high-permeability magnetic shield protects the shielded area.*

uration in a power transformer, contain components of 180 cycles and even higher frequencies as well as 60 cycles. While the magnetic shield may be quite effective against 60 cycles, it is not so effective against the higher-frequency components. So a well-shielded input transformer must utilize another kind of shield to guard against the higher frequencies.

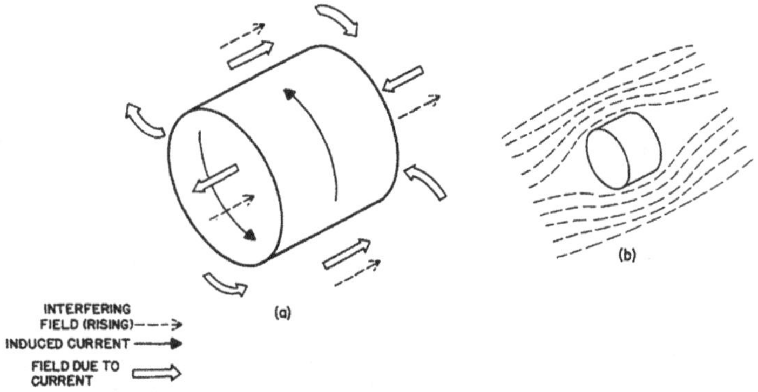

Fig. 506. *Electromagnetic shield.* a—*principle of generating an opposing field.* b—*resultant field pattern.*

This is the electromagnetic shield (Fig. 506). The changing magnetic field induces currents in the conducting material of the shield. These currents in turn induce a field opposing the original field inside the area shielded. At the same time they augment the field outside the area shielded. But the effect depends upon the *changing* of the magnetic field rather than on conducting the magnetic field itself and for this reason is more effective at higher frequencies.

A combination of the two in an arrangement shown in cross-section in Fig. 507 provides an effective shielding for an input transformer.

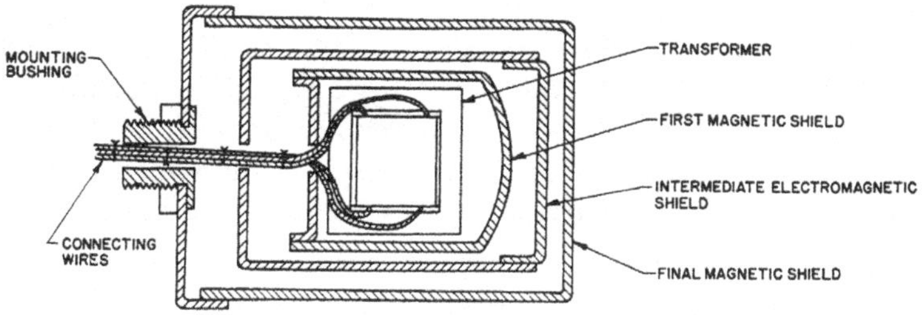

Fig. 507. *Where large amounts of shielding are necessary, a triple arrangement is more effective than one heavy shield.*

Cable microphony

Before we leave the input circuit we must discuss the features that make for cable microphony. The principal cause is change in shield resistance. This means that return currents in the outer conductor or shield of the connecting cable can be modulated by movements that fluctuate the resistance very slightly. This is why body movements of a speaker wearing a lavalier microphone can cause sound to be induced in the cable.

The only way to eliminate this is to use cable that is relatively insusceptible to this effect. The best type is one which uses a twisted rather than braided shield. The braided shield makes a large number of relatively high-resistance contacts between the crossing shield wires. It is also not 100% effective in providing electric shielding because any woven braiding inevitably has air spaces between the wires where they cross.

A twisted shield wrap can be arranged so that all the conductors lay closely side by side and produce a much more effective covering. It might be thought that this arrangement would be inductive but this is prevented by using a single straight conductor inside the shield which short-circuits its successive turns. This it does much more effectively than the braided variety and produces a quieter form of microphone cable.

Also quite effective is the conducting-cotton variety in which the shielding material is cotton made relatively conducting by special treatment. This is wound in a similar manner to the shield just described. This is even more flexible than the wound shield and provides quite effective protection against microphony, if it is well constructed so that the copper connecting wire provides good contact at all points with the conducting cotton.

Input stage noise

Passing from the input circuit, we come to the first stage of amplification. This will add its quota of noise to that already present from the input circuit. In the input stage we have a number of conflicting factors with respect to the choice of tube for producing the lowest resultant noise in the output. The straight triode tube has a noise due to thermal effects given by the equation

$$R_n = \frac{2.5}{g_m}$$

This expresses the noise in terms of an equivalent resistor connected from grid to cathode or grid to ground of the input circuit, g_m being tube transconductance in amperes per volt.

As an example, a 12AX7 tube is quoted as giving a g_m of 1.25 ma per volt (1,250 μmhos) with a bias of −1 volt and plate voltage of 100. This means its equivalent noise resistance will be $\frac{2.5}{.00125} = 2{,}000$ ohms. Note that the g_m at the operative condition *actually used* must be taken, as the value changes for different operating conditions.

If the value of the resistor comes out higher than the resistance of the input circuit, then the noise from the tube will be more than that from the input circuit. If the equivalent resistance produced by the tube is lower than that of the input circuit, then the tube noise will be below that of the input.

If the input impedance is stepped up to the region of 80,000 ohms, any audio tube would have a noise figure well below this so that the controlling feature will be the input source. However, sometimes it is desired to avoid the use of an input transformer, and the source resistance of the pickup, or whatever the input comes from, may be in the region of say 5,000 ohms. In this case the equivalent noise resistance of the input tube may not be very much lower than that of the input circuit itself, so the choice of tube can become important.

If a moving-coil pickup cartridge with a resistance in the region of 100 to 500 ohms is used without any transformer, then the tube will definitely be the limiting factor as to input noise.

Associated with choice of a low-noise tube is the question of gaining sufficient amplification to get the signal above noise that may be introduced in components before the next stage. For example, the plate-coupling resistor has to carry current and the passage of current causes current noise in the resistor. Even though a good brand of resistor for this purpose is chosen, the noise generated will be greater than the thermal noise for the same value resistor. This means that the first stage should have as much gain as possible to lift the signal level above the noise likely to be generated between the first and second stages.

Pentodes give better gain than triodes but suffer from a higher equivalent noise resistance. They have an additional form of noise which occurs because the electron stream inside the tube randomly fluctuates between screen and plate. This random fluctuation produces a noise additional to that normally associated with a tube. The control grid does not cause a similar variety of noise for the reason that its potential is kept negative, hence electrons do not strike the control grid. All the electrons leaving the cathode of a triode tube, under normal operation, proceed to the plate under the influence of the control field from the grid.

But in a pentode, the screen or second grid operates in two ways. Its voltage provides a field that controls the course of the electrons and also, due to the fact that it is positive, attracts some of the electrons to hit it. Some of them will try to reach the screen and just miss, continuing their journey to the plate. This means that a pentode is an inherently more noisy tube than the triode.

This partition noise, as it is called, can be reduced by designing the tube in such a way that the screen grid is in a kind of shadow from the control grid and so receives a small proportion of the total cathode current. But there is a limit to the possible reduction in this manner and the pentode is still more noisy than a triode.

The next best thing, then, would appear to be a high-slope triode. This should yield a low equivalent noise resistance and at the same time magnify the signal sufficiently to be above possible noise generation in the plate circuit and the following stage.

Microphonic tubes

The difficulty here is that a high-magnification triode tube also has a much greater tendency to microphonics. The measures adopted by the tube manufacturers in reducing microphony can be used but they will be less effective on a high-slope unit. Any vibration reaching the electrode structure will be much more effective in generating fluctuating signals than it would on a low-slope tube.

A useful compromise is the cascode circuit (Fig. 508). This virtually uses a twin triode as an amplifier and impedance multiplier to get a gain comparable to a pentode from the two tubes. It

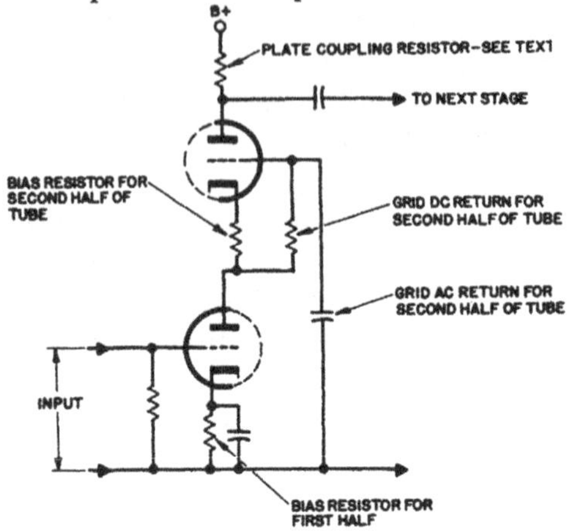

Fig. 508. *Essential components of a cascode input circuit.*

avoids the use of a plate-coupling resistor for the first tube, which feeds directly into the second tube.

The impedance presented by the cathode of the second tube as a plate load for the first depends upon the plate-coupling resistor used in the second. Assume a tube such as the 12AY7, which has an amplification factor of 40 and an average plate resistance of 25,000 ohms. Suppose we use a plate-coupling resistor of 100,000 ohms in the upper tube. The working gain of the upper tube will be 40 multiplied by 100/125, or 32. The grid–cathode swing will be 1/32 times the plate–cathode swing, for the same current swing. As the plate–cathode swing is related to the current swing by the coupling-resistor value, which we have chosen at the moment as 100,000 ohms, the effective resistance seen at the cathode will be 100,000 divided by 32, or about 3,000. This is the load, or coupling "resistor" as it were, for the lower tube.

With an amplification factor of 40 and a plate resistance of 25,000 ohms, the lower tube will now produce a gain of 40 times 3/28, or about 4.3. The overall gain of the double tube thus is 4.3×32, or 137.

This will give plenty of amplification to the input signal by the time it reaches the second plate so that noise in the plate-coupling resistor is not likely to contribute appreciably. However, the noise of the second half tube may contribute an appreciable component because of the low gain achieved by the first, especially as the grid circuit components of the second half may generate a little noise—principally the capacitor coupling the grid to ground.

But increasing the plate-coupling resistor can raise the overall gain of the two stages to a much greater extent than it would with the simple triode stage, using a tube with the same plate resistance.

Assume that we increase the coupling resistor from 100,000 to 500,000 ohms. The gain of the second half of the tube then becomes 40 multiplied by 500/525, or about 38. This is not a very large increase, but the grid–cathode swing is now the plate–cathode swing divided by 38, which means the effective load resistance for the first tube is now 500,000 divided by 38, or approximately 13,000 ohms. This will improve the gain of the lower tube to 40 multiplied by 13/38, or approximately 14, in place of the previous value 4.3.

So increasing the plate-coupling resistor of the upper tube from 100,000 to 500,000 ohms increases the gain of the lower tube from 4.3 to 14 and of the upper tube from 32 to 38—an overall increase from 137 to 532. This means an approximately 10-db improvement in signal–noise ratio at the grid and cathode of the upper tube as well as in its plate circuit. Thus, where the limiting factor is the noise in the tube, a cascode seems to offer the best answer.

The best solution for microphonics in input tubes is to select a type deliberately designed to produce low microphonic effects. If a tube is used in which particular attention has not been paid to this feature, although you may be lucky with one particular tube, you may find that the circuit will seriously deteriorate when the tube is changed and you may have considerable difficulty in obtaining a second satisfactory tube.

Tube hum

Tubes can also be responsible for injecting hum into the circuit. The power supply can be responsible too. See to it that the B-plus supply to each tube is adequately filtered so that no ripple is injected at this point. This can easily be checked by shunting across additional filtering to test whether this is where hum comes from.

There is a variety of other causes of hum originating in tubes. One that should be obvious but sometimes gets overlooked is a satisfactory ground connection to the heaters or filaments. If an ac supply is used for this, a center-point ground is desirable although modern tubes have been so well designed to minimize hum pickup from the heaters that sometimes little difference is noticed so long as the heaters *are* effectively grounded. The kind of hum caused by not grounding the heaters is produced by electric induction due to capacitance transfer from other windings in the power transformer.

Another form of hum is caused by the heater acting as a cath-

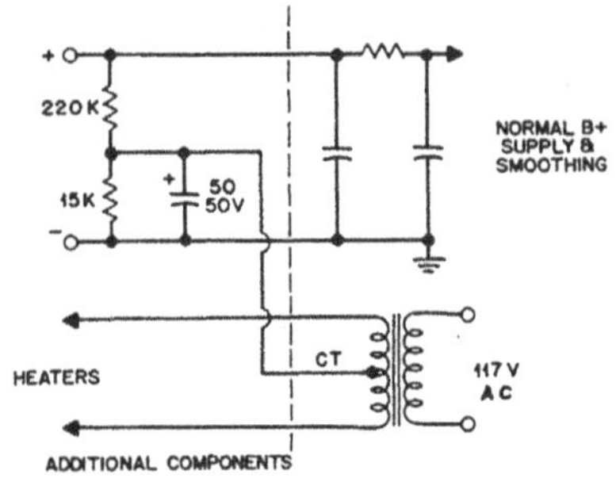

Fig. 509. *Modification to a supply circuit to minimize some kinds of heater hum trouble.*

ode, as well as the true cathode. The cathode–heater insulation prevents the emission from the heater from joining with the normal cathode emission. But the heater usually has a small free space at the ends where it enters and leaves the cathode, and electrons emitted at this point may find their way erratically into the tube electron stream. Since the voltage on the ends of the heater fluctuates at the 60-cycle supply frequency, these stray electrons will be modulated and thus induce a hum in the circuit where the electrons land, probably the plate of the tube.

A way to minimize this is to provide a slightly positive potential on the heater, superimposed on the supply voltage. This should be about 25 volts and can readily be provided (Fig. 509) in the power supply. The heater bias point should be decoupled to ground by a large electrolytic to avoid electrical transfer. This avoids heater emission hum induction by insuring that the heater, instead of acting as an emitter, acts as a secondary plate or diode for some of the electrons emitted by the cathode. While electrons are busy transferring from the cathode to the heater, because the heater is at the more positive potential, there will be no trouble with stray electrons going from the heater to the plate.

There is one more possible cause of hum being induced in an input stage tube. This is a magnetic field that modulates the course of the electrons. If the hum field comes from outside the tube, the simplest cure is to provide the tube with a metal shield. This will usually minimize the effect down to where it cannot produce any noticeable hum injection.

Sometimes, as a precaution against the ac field from the heater, it is advisable to provide a dc supply for the heater. A few years ago this was almost a vital precaution with high-gain amplifiers using very low input. But with modern tube developments, the advantage gained by dc operation is almost negligible. Heater construction now allows for ac operation without appreciable hum field being radiated inside the tube. However some designers still prefer to use dc operation as a final precaution against tubes which might be slightly defective in this regard, while functioning perfectly otherwise.

Distortion

This chapter has concerned itself with the low-level aspect of input stages, by far the most critical one. However dynamic ranges being what they are, sometimes the upper level can also be a consideration. Distortion must be taken care of in the input as well as the output stages of an amplifier.

One does not normally think of a low-level input stage as requiring feedback to reduce distortion because the amplitude of swing is so small that the tube curvature is negligible. Generally speaking, it is not advisable to use feedback over a low-level input stage because it will increase the resultant noise.

Random noise or hiss is the only thing that feedback does not reduce in proportion to the reduction in gain. Any amplifier has a restricted frequency range, hence a phase delay, so it is impossible for the noise to be fed back *exactly* in antiphase with the generated signal. This means that complete cancellation of the noise is impossible with feedback. The noise generated in the feedback amplifier tends to be concentrated at the higher frequencies because cancellation of the lower-frequency components is more effective.

However, the fact that the input stage does not use feedback may mean that it will become the limiting factor for *distortion* because negative feedback can reduce the distortion of the later stages, which would otherwise be much greater. Early stages can produce quite appreciable distortion even with inputs of a few millivolts. In fact, measurements of distortion seem to show that tubes suffer from a sort of "hysteresis" effect like magnetic materials.

The hysteresis of a magnetic material limits minimum distortion to the region of 1% of the magnetizing current. Except at the lowest frequencies, the magnetizing current is a very small fraction of the transfer current. Actual distortion due to an input transformer is far below 0.1%, except at the bottom end. This is a good enough reason why some try to do without an input transformer.

With tubes, the apparent hysteresis effect is always much lower than 1%—usually nearer the region of 0.1%. But the fact that all tubes seem to have a minimum distortion figure when level is reduced suggests that it is well to select a tube type that has a good minimum distortion figure. This information is not generally published—or very often even known. The only way to find out is to make measurements with a good distortion-measuring set. Several samples of each tube type should be tried to get an average and determine what "spread" can be expected.

This is also an argument for achieving as much gain in the first stage as possible so as to get out of this distortion-generation trouble (at higher levels distortion can be reduced more effectively by feedback) as well as getting away from background noise.

Intermediate stages

We also need to consider the intermediate stages between the first input and the phase splitter. Each of these needs to be designed with adequate gain and at the same time with sufficient "headroom" or amplitude-handling margin to avoid running into distortion throughout the dynamic range that the stage has to accommodate.

In a "straight" amplifier where every stage handles a signal larger than the one before it, there is little likelihood of running into distortion before the stage ahead of the phase inverter. However, when a volume or gain control is used in the amplifier—which inevitably has to be put somewhere—it is not impossible for some stage *before* the volume control to be operating at a higher level than the stage following it.

If the control is likely to be worked toward its "bottom end," this feature should receive special attention. If the input of the amplifier is expected to receive a wide range of input levels—from a few millivolts up to perhaps almost a volt—we have conflicting needs to meet. In the interest of achieving a low noise level for the low-level signals, the volume control would be put after the second stage (Fig. 510).

While this would be an optimum position for low-level input, a high-level one will produce so much swing at the second stage as to cause distortion, although the volume control may be quite capable of controlling the *overall* gain so as to give a reasonable output level.

The only way in an amplifier of this type is to provide a

ganged volume control. To get the best results, complimentary tapers should be used in the control so that, for low-level input,

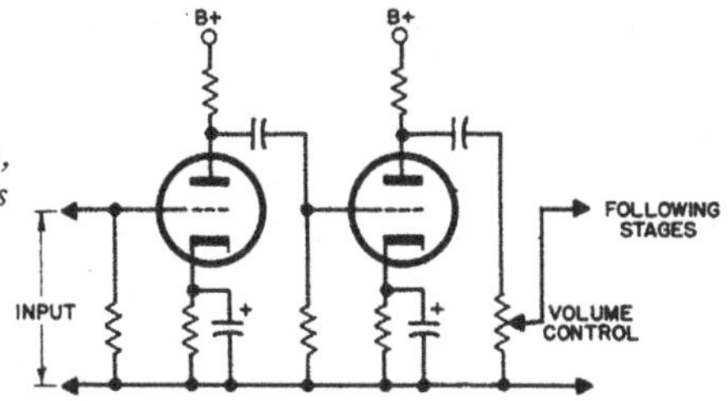

Fig. 510. *For very-low-level inputs, the best place for a volume control is after the second stage or later.*

most of the controlling is done after the second stage or maybe even later in the amplifier, while for high-level inputs most of the controlling is done after the first stage to prevent overloading at any later stage. Fig. 511 shows such a circuit, with the desired tapers for the respective volume controls to achieve a logarithmic control in the combined arrangement.

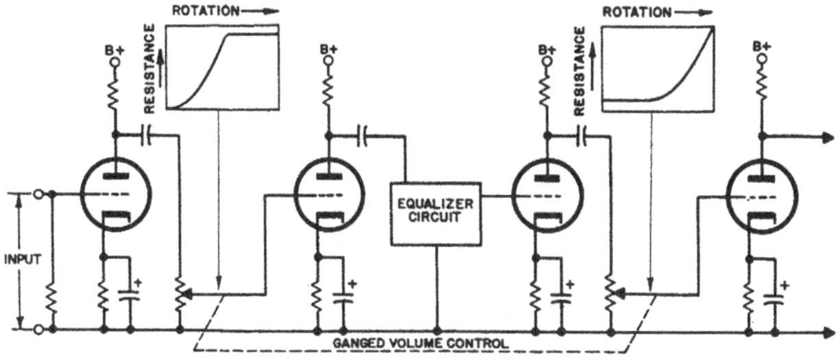

Fig. 511. *A basic arrangement for a preamplifier required to accept a very wide range of input signal.*

While a logarithmic taper, such as used in the "active" position of each potentiometer, is quite customary, a control that is inactive for a portion of its rotation, as each of these are, is not usually a stock item and has to be made specially for such applications. If you are making only one preamp, it is better to design it with separate low- and high-level inputs.

Bootstrap pentode input

There is a particular type that once achieved considerable popularity—using a pentode in conjunction with the split-load type phase inverter (Fig. 512). It is well known that the pentode has a very high plate resistance.

A voltage-amplifying pentode may well have a plate resistance of several megohms with a transconductance between 1 and 1.5 ma per volt. This means that its amplification factor must be several thousand. But such an amplification cannot be achieved in practice because of the necessity for using a reasonable value of plate-

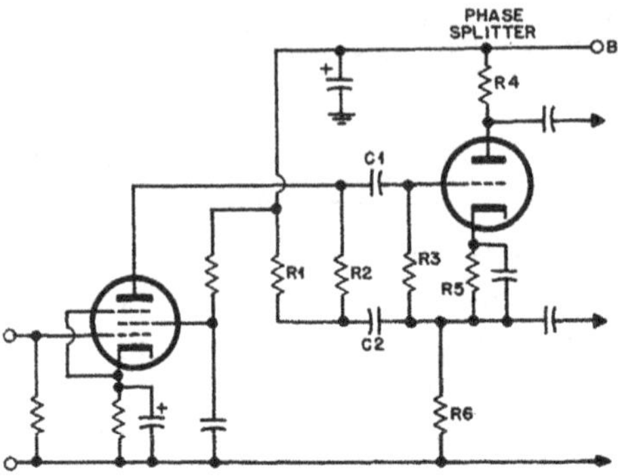

Fig. 512. *Pentode bootstrap input stage.*

coupling resistor. It is not practical to use a coupling resistor of several megohms. Even if a high-voltage supply were used and the tube biased to operate successfully with such a high value resistor, the grid input capacitance of the following stage and stray capacitance would cause excessive high-frequency loss due to the fact that the source resistance at this point would be in the region of a megohm or higher.

The circuit of Fig. 512 makes use of this extremely high pentode gain by using what is actually a normal value of coupling resistor but increasing its *effective* value because it is virtually in the grid-to-cathode circuit of the following stage. Resistor R2 serves as a plate coupler and is virtually in parallel with R3, the actual grid bias resistor, being coupled to it by C1 and C2.

Since the phase-inverter tube has a 50% degeneration, due to R1 in parallel with R6 in its cathode, the effective value of all components connected between grid and cathode is multiplied by the gain of the tube as measured with grid to cathode for input and grid to ground as output. So if R2 is actually 220,000 ohms and R3 is 1 megohm, the combined effective resistance in parallel is about 180,000 ohms. If the phase inverter uses a medium-slope tube with a mu of 40 and a working gain with the resistors used of about 22, this will be a gain, with grid to cathode as input and cathode to ground as output, of 11, or a $(1 + A\beta)$ of 12. So the effect of the 180,000 ohms, from the viewpoint of the pentode plate, will be 12 times 180,000 or over 2 megohms.

With these values the pentode should give a gain in excess of 2,000. The circuit is completed by using R1 in parallel with R6 as the cathode load, since they are coupled together by C2 at the top end, and by R1 returning to B plus, which is decoupled to ground, at the bottom end. So, to get true balance in the phase inverter, the combined parallel resistance of R1 and R6 must be equal to R4 in the plate.

The high-frequency response is maintained because the *effective* grid-to-cathode capacitance of the tube is also divided by the same factor that the effective source resistance is multiplied by. So the high-frequency rolloff will be at the same point that would be produced if a 220,000-ohm resistor were used as plate coupling in the normal manner, with a 1-megohm grid resistor for the following stage.

Provided this circuit is used with low-level input—which it would have to be with such phenomenal gain—the distortion is fairly low although operation of a pentode with such extremely high load resistance tends to produce distortion even at quite low levels.

The application of overall feedback in any quantity becomes somewhat of a problem with this circuit because of the extremely high impedances involved and a necessity for careful choice of coupling capacitors for C1 and C2. If these are made too large, they will result in excessive high-frequency rolloff due to their capacitance to ground. If they are not large enough, they result in partial removal of the multiplying effect produced by the feedback at the low-frequency end, hence result in the pentode stage losing gain down at these frequencies.

Capacitor C2 effectively provides a form of positive feedback, from the viewpoint of the pentode, although from the viewpoint of the phase splitter it is part of the negative feedback arrangement due to degeneration in the cathode.

The screen decoupling of the pentode adds a further complication with this method of operation. Under normal conditions removing decoupling of a pentode stage changes the effective gain of the stage, approximately from that as a pentode to a triode. The effect of the decoupling capacitor is to introduce a step response at some low frequency, producing a phase shift similar to a rolloff but without the same degree of attenuation. It has a similar effect here except that the gain as a pentode is very much more than usual so the effect of deterioration in decoupling at the screen at low frequencies will reduce gain by a factor approaching to 100 to 1. This is almost the same as another low-frequency rolloff.

For a similar reason, if the decoupling circuit is at all inductive, it will produce deterioration of response at the high-frequency end due to the necessity for complete screen decoupling to achieve the high pentode gain realized.

Transistors

The use of transistors has provided another possible form of input stage and it is one that holds considerable promise of proving ultimately to be the best. When transistors were first introduced, they seemed to suffer from serious signal-to-noise problems but this was because they were not used under the best operating conditions. An advantage of the transistor circuit is that its input impedance can be made to suit the low input resistances normally used for the best microphones and pickups, without the need for a matching transformer. By careful choice of operating conditions, the distortion can be kept quite low and by working with a very low collector current the noise level can be reduced to a value that compares favorably with the best tube circuit. Also transistors are inherently free from hum and microphony troubles that are always present in some degree with tubes.

chapter 6

matching

IMPEDANCE matching is necessary in audio circuits. There are two principal places for matching: at the input and output of an amplifier. In each, the purpose is to make an *actual* impedance look like the impedance *required* by the amplifier, as an optimum input or output impedance.

Input Circuits

In input circuits, matching arrangements change the actual input impedance to a value that will give the best signal–to–noise ratio compatible with a suitable frequency response. As an example, a 50-ohm microphone working directly into an amplifier will not have a very good signal–to–noise ratio because the equivalent noise resistance of the tube will be 1,000 ohms or more, about 13-db higher in noise level than that in the source itself.

A transformer to step up the signal to the grid of the first tube will improve this situation. It will also contribute some useful gain so that a smaller amount of gain is required in the amplifier. This might solve some feedback problems. It is more difficult to get, say, 20-db feedback over an amplifier with higher gain than one with lower gain because the higher gain necessitates more stages and complicates the design. However, the use of too much stepup in the input transformer will deteriorate the frequency response—quite apart from that of the transformer itself.

Assuming that the transformer steps up the 50 ohms to 80,000 ohms, using a 40 to 1 stepup, then the microphone now looks like an 80,000-ohm resistor as far as the grid of the first stage is concerned. If the input capacitance should be as high as, say, 100 μμf (which it might well be) including the secondary capacitance of the transformer, then the response will be 3 db down at 20 kc because 100 $\mu\mu$f gives a reactance of 80,000 ohms at 20 kc. If twice the stepup were used, (resulting in 320,000 ohms) the same secondary capacitance would cause the response to be 3db down at about 5 kc.

Thus, the choice of stepup is a compromise between signal–to–noise level and the frequency response required for a full range within close tolerance. Probably a 20-to-1 stepup with the 3-db point at 80 kc will be considered an acceptable ratio.

The input transformer does not directly affect the low-frequency response unless the primary inductance is inadequate. However, there is a way in which the amount of stepup does sometimes affect the low-frequency end. For a satisfactory high-frequency end with low leakage inductance and secondary capacitance, the winding has to be physically small. This means the transformer itself must be physically small. There is a limit to the fineness of wire with which the secondary winding can be wound, consequently there is a limit to the number of turns that can be used. The only way to increase step-up, when the limit of turns has been reached in the secondary is to reduce the number of turns on the primary, which of necessity also reduces the primary inductance.

Inductive sources

We have assumed the source impedance to be a pure resistance of 50 ohms. Many magnetic microphones and phono cartridges, however, possess quite an appreciable inductive component although the nominal impedance may be given as the approximate *resistance* presented at a middle frequency, somewhere from 400 to 1,000 cycles.

Where the transducer (microphone or pickup) possesses appreciable inductance, it will be considerably greater than any leakage inductance that an input transformer will add. An input transformer designed for use with an inductive device does not need to have particularly low leakage inductance because the inductance of the source impedance will swamp it anyway. But this also means that a smaller capacitance can be tolerated on the secondary before the high-frequency-rolloff condition becomes severe and, therefore, that the effective stepup is correspondingly restricted.

In some instances it may be difficult to prevent the capacitance on the secondary of the transformer from resonating with the inductance of the source impedance. In this case the response will show a peak at the high-frequency end. To damp this peak without risking deterioration of the low-frequency response apply resistance across the *secondary* of the input transformer (Fig. 601).

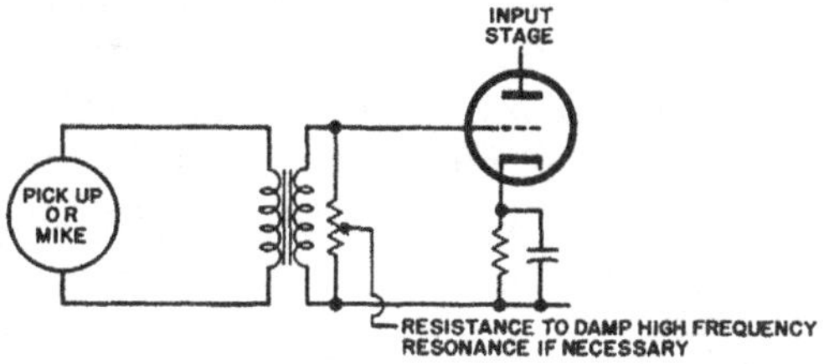

Fig. 601. *Controlling high-frequency response through an input transformer by using a loading resistance on its secondary.*

Try different values until the optimally flat response is achieved. If this results in a rolloff at a frequency lower than is considered satisfactory, then the only solution will be to use a transformer with less stepup.

Capacitive sources

Input matching also raises the question of suitable connections for capacitor and piezoelectric microphones and pickups. Here our thinking has to take into account the fact that the source impedance is no longer a resistance or a resistance combined with inductance but principally a capacitance (Fig. 602).

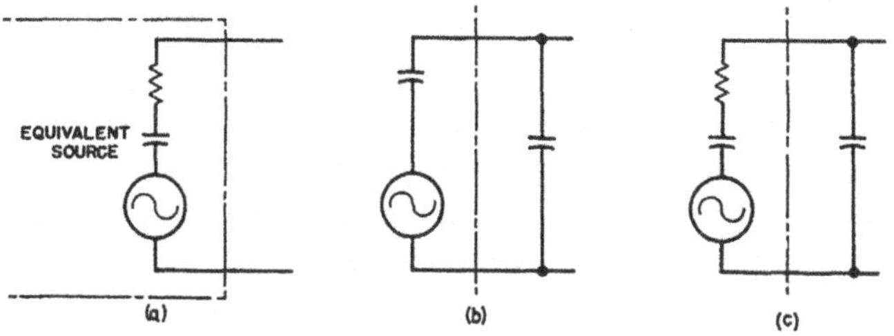

Fig. 602. *Matching as applied to capacitor or piezoelectric transducers.* a—*true equivalent source.* b—*an ideal matching arrangement, assuming the source is entirely capacitive, giving response from zero to infinity.* c—*the resistance in the actual source prevents the ideal of* b *from being realized.*

Working from a capacitive source into capacitive load will result in a flat frequency response from zero to infinity, in theory. But there are always some resistance elements somewhere in the circuit to introduce losses and these will result in a rolloff somewhere in the frequency response.

The source usually looks like a capacitance with a little bit of resistance in series, due to the losses in the device and the dynamic loading. This results in a high-frequency rolloff. If the transducer is worked into a capacitance larger than itself, the high-frequency rolloff will be determined by the transducer capacitance and if into a capacitance smaller than itself, then the loading capacitance will determine the high-frequency rolloff point. Usually a resistance provides a grid circuit return for the input tube and this resistance (Fig. 603) results in a low-frequency rolloff.

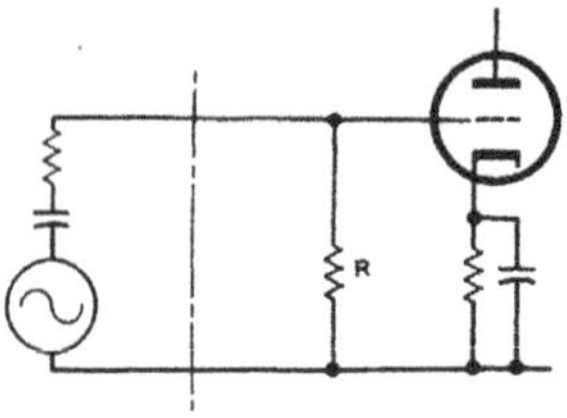

Fig. 603. *Loading a capacitive source with a resistor R produces a low-frequency rolloff, the frequency of which will be determined by relative values.*

Optimum energy transfer from the source to the load occurs if the source and load capacitances are equal. This seldom happens but it is not feasible to use a transformer for matching here because the inductance and losses of the transformer result in impossible frequency response limitations before anything approaching ideal matching can be achieved. In practice, it is impossible to design a transformer that will achieve any improvement on the direct connection.

There is a particular application with the piezoelectric type of microphone or pickup that is sometimes useful. The output from these devices is relatively high compared to other transducers. Consequently it may be too large in some instances for the input stage to which it will be connected. Also, to achieve a satisfactory low-frequency response a very large input resistor is required. But, by employing a large input capacitance (compared with the value of the source) a more suitable input resistance can be used without

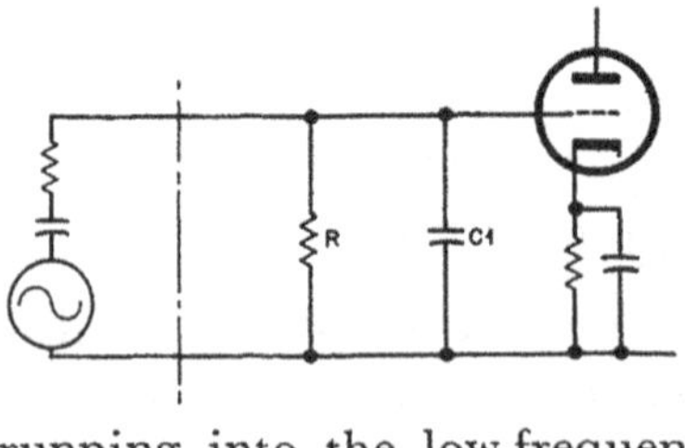

Fig. 604. *In cases where the transducer gives plenty, if not too much output, C1 is used to attenuate the output, so a lower value dc grid-return resistor R, can be used without undue low-frequency loss.*

running into the low-frequency rolloff. Also the input will be attenuated into a region more suited for the amplifier (Fig. 604).

To maintain the high-frequency response which, under these

circumstances, would be limited by the relation between resistance and capacitance in the transducer itself, it may be necessary to add a component of resistance in series with the artificial input capacitance to maintain level response at the high end (Fig. 605).

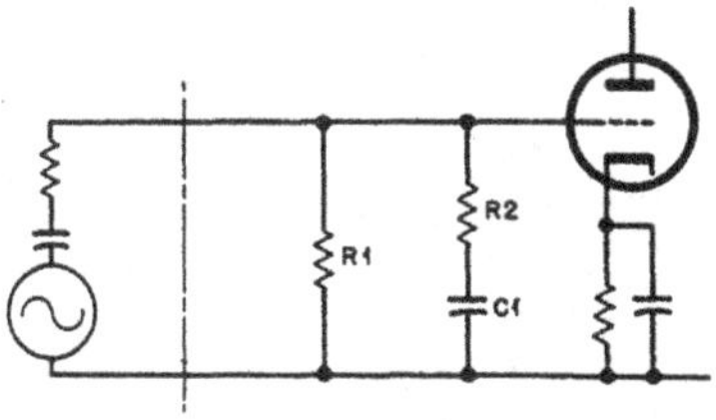

Fig. 605. *If the arrangement of Fig. 604 results in high-frequency loss, due to the equivalent resistance in the source, it can be offset by a suitable value of resistance, R2.*

This would be a useful circuit for permitting a piezoelectric pickup, for example, to be connected to the input of a preamplifier intended for a magnetic or moving-coil type pickup. The only thing left here, however, is the fact that the crystal pickup is a constant-amplitude device while the magnetic or moving coil are velocity devices. There should be a 6-db-per-octave upward slope in the input circuit from the piezo pickup for the equalization provided in the amplifier for the magnetic or moving-coil cartridge to be suitable for the crystal type. To achieve this all that is necessary is a considerably lower value of loading resistor on the input (in the circuit of Fig. 603) which produces a 6-db-per-octave bass loss from some relatively high frequency. This will amount to some 35 or 40 db attenuation at 1,000 cps. For this reason it will not be necessary to use the capacitive attenuation shown in Fig. 604.

With some capacitive transducers, such as the capacitor microphone or the sound-cell crystal, the sensitivity is very poor and so it is necessary to make the best of the output available without introducing this attenuation effect. To achieve this a very high input resistor is required in the first grid. This will result in a relatively high background noise. A method of partially overcoming this is to use some feedback over the input stage that will increase the effective input impedance, rather like a cathode follower.

The arrangement could be a cathode follower to act as an impedance changer or it could be part of the overall feedback arrangement (Figs. 606 and 607). In these circuits the actual grid–to–cathode resistance can be much lower, utilizing the advantage of a lower input noise, but the effective value of resistance from the point of view of loading the transducer is increased by the degeneration. This also decreases the capacitive loading presented by the tube input, which for these transducer types may well be equal to or greater than that of the source capacitance. The degeneration, as well as increasing the effective load resistance, will also increase the noise to some extent. But degeneration does at-

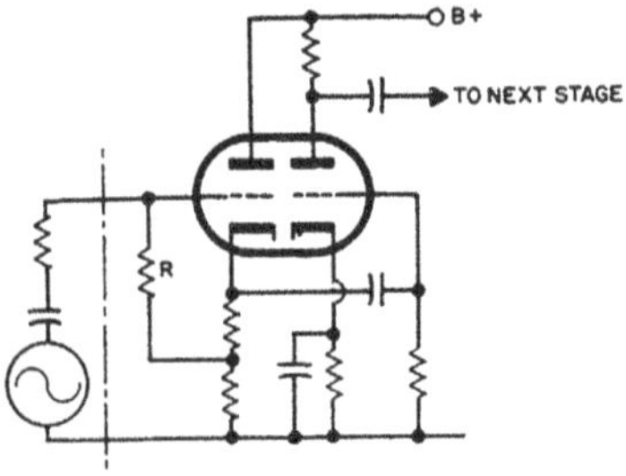

Fig. 606. *Use of a complete cathode follower at the input to raise the effective value of R for loading a capacitive source, without a corresponding noise increase.*

tempt to cancel some of the noise voltages generated although it is not completely successful. So the overall result is an improvement over using the actual resistance instead of the effective resistance produced by this degeneration.

Output circuits

The usual reason for matching in output stages is the achieving of maximum output power for given operating conditions in each of the methods of tube connection. Whatever the reason for matching the optimum load resistance is invariably different from the plate resistance or effective source resistance of the stage.

For maximum efficiency as a power transfer device, the load

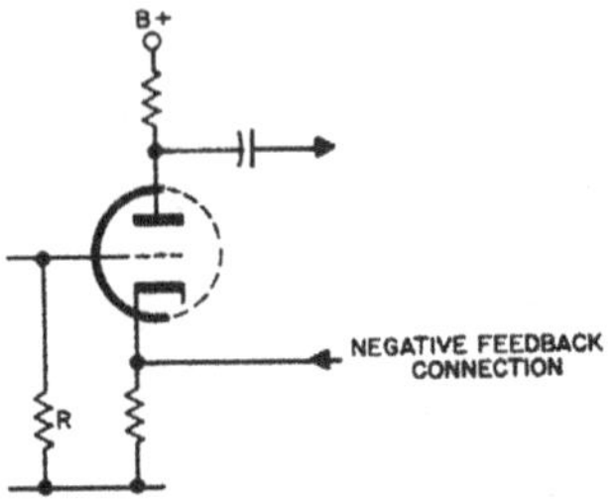

Fig. 607. *An alternative way of helping raise the effective value of resistor R is by negative feedback from a later stage, connected to the cathode.*

resistance or impedance should be equal to the source resistance or impedance. This is based on a theoretical linear device without any limitations. If tubes satisfy this condition, then the load resistance should always equal the source resistance. But unfortunately the value of source resistance, or plate resistance, of the tubes is only applicable over a certain range due to the curvature of the tubes and it is desirable to steer the load line away from this curvature to maintain linear operating conditions to the greatest possible extent.

Maximum power output

This is why triode operation, with curvature at the bottom of the line, requires a resistance higher than the working plate resistance so as to keep the load line more horizontal and away from the curvature. (Except, of course, when low loading is used, in which case the curvature of the two tubes in push-pull is used to cancel second-harmonic distortion and provide a bigger output.)

In the case of pentodes, the curvature is toward the top or left-hand end of the tube characteristics. A more vertical load line than the slope of the characteristics proves satisfactory to avoid the curvature at the left-hand end.

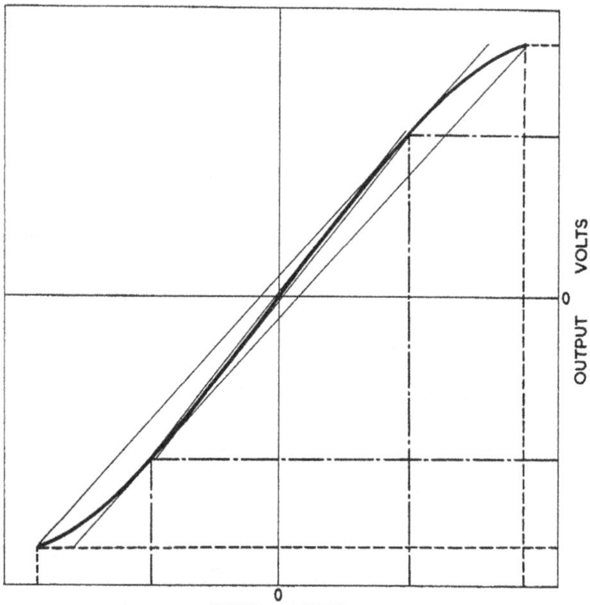

Fig. 608. *A transfer characteristic with "one-way" curvature results in lower distortion at lower levels. Input and output swing at maximum level is indicated by dashed lines, at lower level by dot and dash lines.*

Minimum distortion

This is not the only condition to be considered for deciding correct matching in an output stage, however. Closely associated with the load value for maximum power is the value for lowest possible distortion. Distortion is usually considered on the basis of maximum output. However, distortion at maximum output is less noticeable than distortion at somewhat lower output. Consequently, amplifiers are sometimes designed to give the lowest possible distortion at relatively low levels, allowing the distortion to become a little higher at maximum output.

This is another reason why the circuits which produce mainly low-order distortion are better than those that produce high-order distortion. (Do not confuse *order* of distortion which indicates whether the harmonics caused are *high* or *low,* with *amount* of distortion, given as a percentage figure.) The circuits that produce lower-order distortion use tube characteristics with a simple curvature resulting in either second or third harmonic. Consequently the distortion reduces as the length of the load line is reduced, representing lower levels, and the order of distortion remains the same (Fig. 608). Also with this kind of curvature the slope of the load line, representing its impedance value, remains

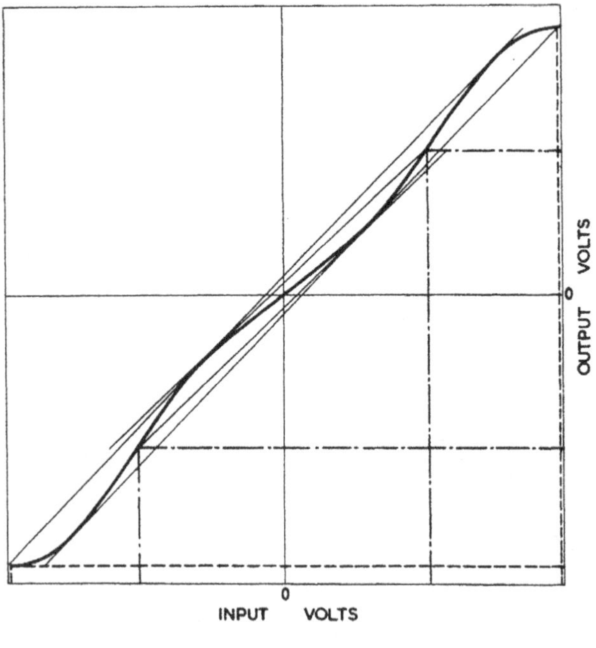

Fig. 609. *A transfer characteristic with several reversals of curvature may result in lower magnitude of distortion (but higher order) at maximum level (dashed lines); but higher magnitude (lower order) at lower levels (dot and dash lines).*

very close to the same angle for minimum distortion at different levels.

But the kind of curvature given by pentode operation, which produces recognizable quantities of fifth and also seventh harmonic, results from a transfer characteristic that has multiple curves (Fig. 609). The distortion at full amplitude contains components of fifth and seventh harmonic, due to the multiple curvature. But as the excursion is reduced, the curvature simplifies and so the order of distortion is restricted to the lower harmonics. However, the magnitude of the distortion may actually increase under these conditions at lower levels, since the multiple curvature reduces the *effective* components of low order at high amplitude.

This cancellation does not occur at low amplitude and we may be left with a higher percentage of low-order harmonic distortion. Readjustment of the loading condition may result in improved distortion at low levels although it will aggravate it at high levels. This is another good reason for avoiding straight pentode operation.

A final feature of matching is the effect of feedback as a source or load modifier. For example, to give a constant output, the electrostatic transducer should have a constant voltage. This means that the source impedance should look like a capacitance or else some means should be undertaken to provide constant voltage across the variable reactance presented by a capacitance load.

For this reason the power amplifier feeding an electrostatic transducer must be designed on the basis of much larger power output—or perhaps we should say volt–amp output (because the load is almost entirely capacitive reactance) than would be necessary for the regular type of transducer.

Transistors

When a transistor is operated in grounded emitter connection, it behaves very similarly to a tube. It amplifies both voltage and current, and produces a phase inversion. Practically all the present circuits for application in audio use grounded emitter operation. One reason for this may be that it avoids too many new concepts by enabling the transistor to be visualized as a "tube replacement." However an analysis of other methods of connection also seems to show definite performance advantages.

The transistor is basically a current-operated device, whereas the tube is voltage-operated. This is important because each kind of amplifying unit has its nonlinearity, or curvature. In a tube, the departure from linear amplification is related to voltages handled. In a transistor, it must be related to the currents.

This fact is reflected in a consideration of matching for transistor circuits. For tube circuits, we think of a voltage generator with a series resistance as our source of input. For transistor circuits it is more convenient to translate this into a current generator with a shunt conductance or susceptance. The important quantity delivered to the transistor is not the voltage fluctuation, but the *current* fluctuation.

This is complicated by the necessity of providing the correct bias for the transistor. The operating temperature of tubes is not a critical factor, and is principally determined by the heater dissipation. So long as electrons are emitted by the cathode in adequate quantities, the temperature has little effect. But in a transistor, change in ambient temperature alters the characteristics considerably. The circuit must be designed so that change in temperature does not interfere with operation.

There are two approaches to this problem: one is to compensate for the known temperature characteristics of the transistor, using diodes of similar material, or temperature-sensitive resistors. What would seem to be the better way uses a dc feedback circuit that adjusts relative voltages in the base-emitter circuit to obtain the right operating range of collector current (and thus base current). This has the advantage that it will also care for deviation in individual transistors as well as temperature effects.

So, particularly in a transistor input circuit, the ac susceptance or impedance must be right to give the required frequency response and to minimize distortion, while the dc conditions must maintain correct operation of the transistor. This subject becomes quite an involved one, and the number of ways of doing it promises to exceed the variations in tube circuits; so it cannot be treated in detail here, and is definitely beyond the scope of this book.

chapter 7

equalization

THUS far we have assumed that the desirable frequency response is absolutely flat—that all frequencies in the audio spectrum must be reproduced uniformly. Now we come to cases where correction has to be made for various nonuniformities that come to light in audio systems.

Why equalizers are needed

For example, due to its own mechanical frequency response, a disc cutter head may not give an exactly uniform cut at all frequencies for the electrical power supplied to it. Equalization is necessary so that, where its response is weak mechanically, more electrical power is supplied to it. The result is that a uniform amplitude of cut is made at all frequencies for a uniform electrical input to the system (Fig. 701).

Because it has no moving parts, a tape recording head is inherently much more linear in its frequency response than a disc cutter head. However, its dimensions control the magnetic field it produces in the tape and this, together with the velocity of the tape and losses in the magnetic core of the head, affect the possible frequency response.

Disc pickups and tape playback heads also produce their own individual frequency responses. The ideal, of course, is to select a pickup that has (as nearly as possible) a flat frequency response and does not need equalization for its mechanical characteristics. However, if a pickup is as good as can be made but a *perfectly* flat response is needed for some reason, it may be necessary to equalize for the slight deficiencies that remain in the pickup.

With a playback head there is definite need for equalization in all cases because the frequency response is rather like that shown in Fig. 702. This can be explained in terms of the overall characteristic of the magnetic recording system.

If a constant current, at different frequencies, is fed to the recording head, the amplitude of magnetization on the tape will be

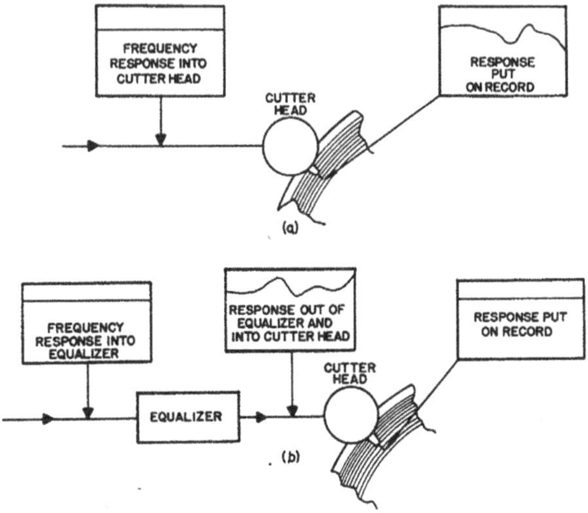

Fig. 701. *Equalizer compensates for the response deviation of a cutter head.*

constant. This means that the maximum amplitude, just where saturation commences, will correspond with the same current in the recording head at all frequencies. This assumes that the losses in the recording head are uniform. In practice they are not, so these will need to be equalized in recording.

On playback, the output from the playback head is proportional to the rate at which the magnetic field *fluctuates* or to the intensity of magnetization at any particular frequency. But with different frequencies it will be proportional to the frequency, assuming the same intensity. Thus the playback head will produce a response with a 6-db-per-octave rise because doubling the frequency doubles the rate at which the field fluctuates.

However there is limit to this response, due to the physical dimensions of the air gap in the head. When these become comparable with the wavelength of the magnetization along the tape, high-frequency loss sets in. So the overall characteristic (Fig. 702)

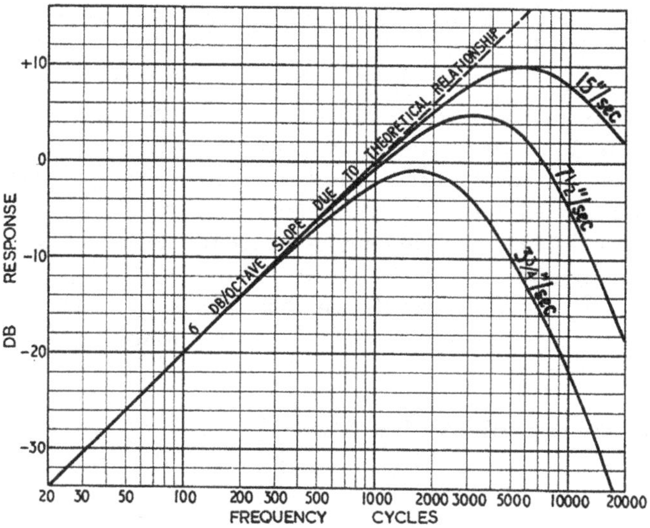

Fig. 702. *Typical response curves of a tape playback head, using a constant-magnetic-density tape at different speeds.*

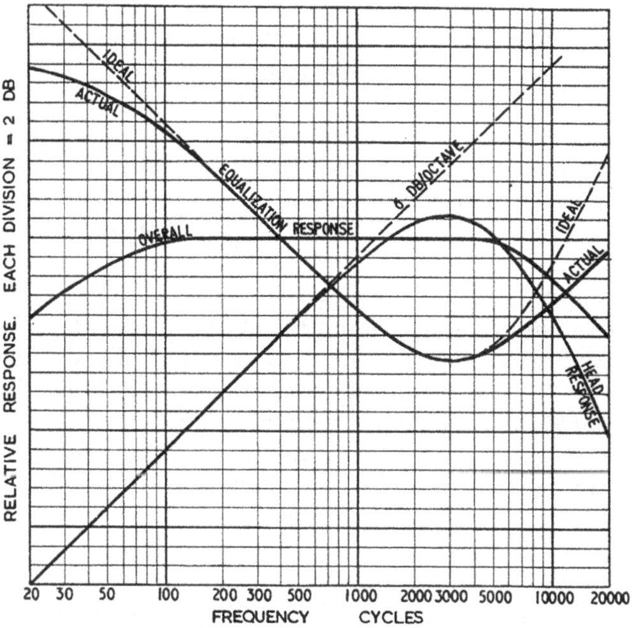

Fig. 703. *Playback equalization for a flat overall frequency response over an acceptable range.*

is a rising response at 6 db per octave until the high-frequency loss begins to pull it down. It reaches a maximum somewhere between 2,000 and 5,000 cycles (according to individual head design and the speed of tape used) and then falls off very sharply.

To get a response to a higher frequency than that naturally produced by the head, equalization is again necessary (Fig. 703).

Equalization may also be necessary for microphones or speakers, each of which can have deficiencies in frequency response which could be compensated for electrically. However, the "progress of the art", as it is called, has shown that this is not really a good policy. The best approach is to design a microphone and speaker to have a response as flat as possible.

Equalization not always the answer

The author recalls a microphone in which the approach used was to eliminate all resonances except one major one. This was achieved by cutting holes in the framework so that there was only one large cavity which was tuned by the mechanical mass of the diaphragm. The result was a microphone with a single peak, about 30 db high, in the region of 500 cycles. This microphone was operated with a built-in equalizer that produced a corresponding "hole" in the response at the same frequency.

Its overall response turned out to be the closest to flat that had been achieved in any dynamic type microphone until that date. From the viewpoint of frequency response alone, it would still be an extremely good microphone. However, its performance left something to be desired, since the electrical equalization could not compensate for the deficiencies of the resonance in handling transients.

This can be illustrated by Fig. 704. When a 500-cycle tone first strikes the microphone mechanically, it will take time to build up to its maximum amplitude, due to the high Q of the mechanical resonant circuit—about 30. This means it will take about 30 cycles to reach 0.637 of its ultimate amplitude. Similar buildup takes place in the equalization circuit, due to the high electrical Q. It will take about 30 cycles of the 500-cycle tone to get 0.637 of the ultimate absorption effect of the circuit.

Now if the pulse stimulus came at a point between the microphone and equalizer, the two would neutralize and produce an output that is a replica of the input. But the fact that the feed starts from the mechanical side of the microphone and goes through that resonance and then the output from that is fed to the input of the electrical equalizer, means that the electrical equalizer does not have the same square-frontage pulse to handle that the microphone does. For this reason the absorption effect of the electrical equalization cannot start operating until the mechanical resonance of the microphone has started to build up.

The overall output will show a kind of ring on the front edge

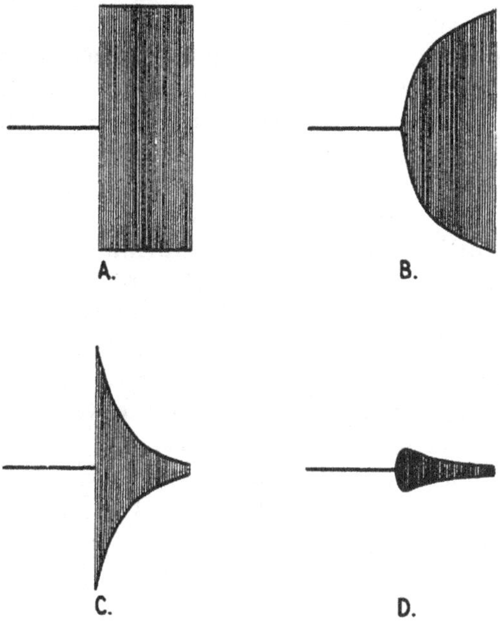

Fig. 704. *Transducers needing excessive equalization should be avoided: a—acoustic input wave to the microphone discussed in text; b—electrical response to wave* a; c—*response of equalizer network to wave* a; d—*overall response, which does not restore shape of* a.

of the 500-cycle tone before electrical equalization manages to control it. Of course a square pulse of 500-cycle tones is not the kind of signal that happens very often in program material. But other kinds of transients are, and the effect of this arrangement is to color any transient with a kind of 500-cycle ring effect.

The sound of a door shutting, picked up by this microphone, is as if all the panels in the door, as well as most of the panels in the room, were tuned to mechanical resonance at 500 cycles. If you don't hear the noise of the door live—only as reproduced over the microphone, this does not perhaps sound impossible but, when other sounds all seem to have the same resonance, it gives a general coloration to the reproduction. This is quite noticeable even though the overall response of the microphone is perfectly flat when a continuous tone of different frequencies is used for making the measurements.

For this reason all kinds of transducers; microphones, loudspeakers, cutters and pickups are, as far as possible, designed to have a flat frequency response in themselves so as not to need electrical equalization.

Another place where electrical equalization may be necessary is in radio transmission and reception. Here, again, the ideal arrangement is to arrange for the various radio circuits to have a flat-topped bandpass characteristic so that demodulation of the program material gives a uniform frequency response without the necessity for any equalization. However, the practical consideration of reducing noise caused by various forms of interference sometimes dictates that the radio response has to be rolled off somewhat toward the extremes of the passband.

This can sometimes be helped by using electrical equalization to bring back some of the lost highs. However, unless a radio circuit is used that discriminates against the various noise or interference components (as well as losing some of the higher frequencies) restoration of the high frequencies in the audio circuit will equally restore the interference! This discussion, however, is somewhat outside the scope of an audio book.

Phono and tape equalization

Another cause for equalization—and in modern audio practice the principal one—is to achieve satisfactory operating conditions for the recording medium used, whether disc or tape.

Taking disc first and using lateral cut (common practice these days), there are two bases for standardization: the *amplitude* of the cut—that is, the width between the maximum excursion from side to side of the groove (Fig. 705), and the *velocity* of the cut—the maximum rate at which the stylus moves in cutting the groove.

As the majority of pickups utilize a velocity principle (the electrical output from the pickup is proportional to the *rate* at which stylus moves) this standard of reference holds most favor. It is a useful reference also because it yields a convenient method of calibrating the grooves actually cut on the disc. From these considerations it would be convenient to use a *constant-velocity* cut in which the same electrical input would produce the same maximum velocity of stylus movement at all frequencies. But this will mean that the amplitude of excursion will be inversely proportional to frequency. See Fig. 706.

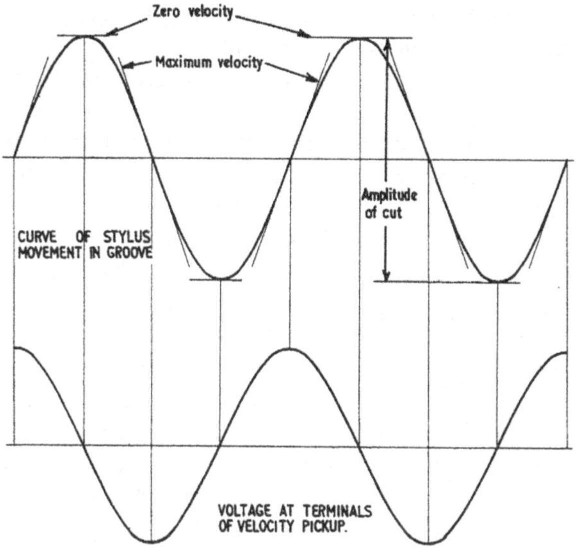

Fig. 705. *Essential features of stylus movement in a groove for velocity type pickups and recording heads.*

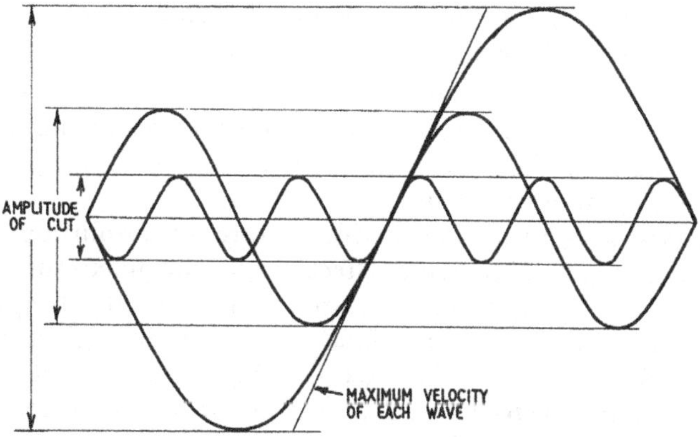

Fig. 706. *Grooves of different frequencies are superimposed to show relation between velocity and amplitude.*

To use some practical figures, if the excursion for maximum output at 20,000 cycles is 1/10,000 inch—a very small excursion—the movement at 20 cycles for the same electrical output would be 1,000 times this, or 1/10 inch. Obviously it is not possible to pack in 100 grooves to the inch (or even more) if *each groove* is going to make an excursion approaching 1/10 inch. The grooves would overlap several times over.

So, in the interest of economy in groove spacing, it is necessary to limit the excursion at the lower frequencies. This is the reason for the low-frequency rolloff in recording. By limiting the constant-velocity principle to, say, 500 cycles and then changing below this frequency to a constant amplitude, the maximum excursion, using the same hypothesis, is restricted to about 1/400 inch. On this basis it will be readily possible to exceed 100 grooves per inch.

But that is not all there is to the problem. As with input stages, we have a dynamic-range problem and this hits us at the high-frequency end. Using the same assumed basis of a maximum amplitude of 1/10,000 inch at 20,000 cycles, this will become 1/5,000 inch at 10,000 cycles (a more usual maximum upper limit although some recordings maintain appreciable output up to 15,000 cycles). Taking the maximum excursion as 1/5,000 inch and the dynamic range as 60 db, the minimum excursion at 10,000 cycles will be 1/5,000,000 inch. This is so very small that noise will drown out signal in this region.

For this reason it is desirable to increase the velocity or excursion of the stylus at higher frequencies. To assist in this, a reversal from the constant-velocity principle to constant amplitude again takes place in the region of 2,000 cycles. For the RIAA curve this changeover takes place at 2,120 cycles.

Different authorities have come to slightly different conclusions as to which is the best compromise in achieving a good working characteristic, but the principles governing the decision are the same in all instances.

Using tape as a recording medium, constant amplitude is the obvious limitation. Saturation effect limits the maximum amplitude that can be put on the tape at any specific frequency. The principal modifications made to the recording characteristic on tape are due to the excessive high-frequency losses inevitable in any playback head and also the inherent restricted dynamic range of tape. To get an effectively low background noise, it is desirable to operate with as little "headroom" as possible between the maximum signal and saturation point at all frequencies.

As most of the energy is concentrated in the mid-range and most of the noise in the higher frequencies, it is good to pre-emphasize the higher frequencies so that their energy level is nearer that of the mid-range on recording. This gives them a better margin above the background noise level in this region

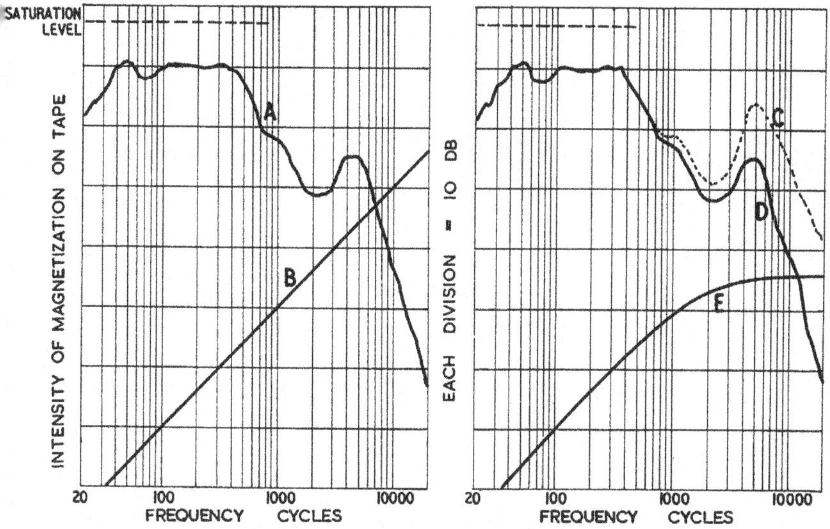

Fig. 707. *How pre-emphasis helps: Curve A is a typical energy distribution curve for program material; curve B is the noise energy distribution; curve C shows energy distribution in program with pre-emphasis. De-emphasis on playback gives curve D, which is the same as curve A, but noise changes to curve E.*

(Fig. 707). At the same time this will help overcome high-frequency loss in the playback head.

Standards

Where an entire system is being used by the same individual for recording and playback, it is not very important whether the exact *record* characteristic and *playback* characteristic conform to any particular response, provided the *overall* response from the input to the recorder to the output from the playback is flat. A variation of a few db in the response on the actual recording does not matter, provided the overall result comes out right. But when recordings are made on one recorder and played back on another, it is highly desirable to have a standard of recording and playback characteristics so that any recorder will produce a recording which, reproduced on any playback system, will come out flat. This is the reason for establishing standards in both disc and tape recording.

While it is easy to say what the standard response *should* be, the question arises of how we can establish that a recording satisfactorily follows a specified response.

In the case of disc recording, this is relatively simple. A visual inspection under a point source of light will readily show this (Fig. 708). If the grooves are unmodulated, parallel concentric

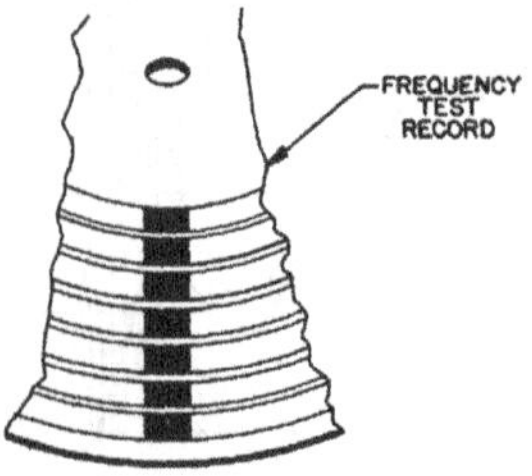

Fig. 708. *Light pattern seen across a test-frequency disc, when each frequency band is recorded at the same velocity.*

grooves, the point source of light will give a narrow line down the record as it reflects in each consecutive groove. But when the grooves are modulated, the angle (due to the velocity of modulation in the middle of excursion) causes an extended band of reflected light (Fig. 709).

So the width of a band of light reflected from a point source depends on the maximum velocity of modulation at this point. This forms a very convenient means of examining a recording and estimating the response with considerable precision.

Fig. 709. a—*with an unmodulated groove, light is reflected from just one point.* b—*when groove is modulated, reflection occurs over an angle limited by the angle between the maximum velocity points in opposite directions.*

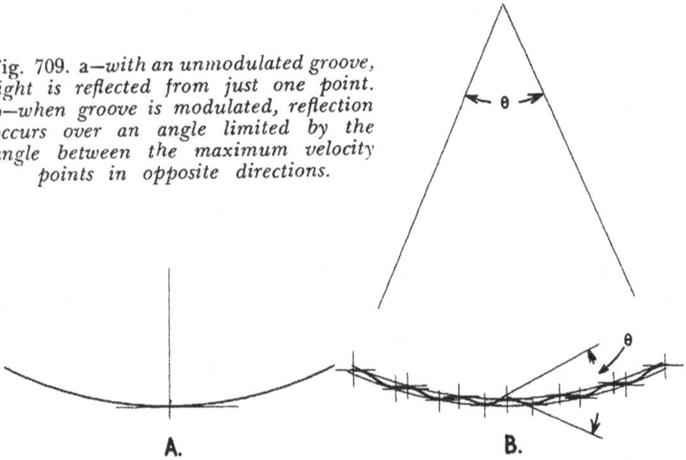

A standard-frequency record can be readily made for testing pickups and record players by recording bands at different frequencies and adjusting the amplitude of recording so the response is uniform or follows the desired playback characteristic. If the cutter head is not accurate, adjustments can be made experimentally to produce a uniform resultant recording in spite of the variation in integrity of the cutter.

Disc "standards"

This sounds so simple that it must be too good to be true! This proves to be the case at the high-frequency end. The same pickups tested on different recordings, each of which look identical under this method of inspection, will give quite different frequency response characteristics. The results may even be contradictory.

This is so because the performance of a pickup depends, not only on the mass and compliance and various mechanical features of the pickup itself, but also on *the compliance of the record material* where the stylus rides in the groove. When the record should move the stylus very rapidly at high frequencies, the groove walls "give" a little bit instead of moving the stylus as they should. The amount by which the walls give will vary according to the characteristics of individual pickups and the material of which the disc is made.

The characteristics we see when we examine the record under a point-source light beam only indicate what the grooves are when they are static. It does not indicate the shape they take when a stylus rides in them and, of course, that shape may vary according to the pickup to which the stylus is fixed.

This situation in disc recording is still in a somewhat unsatisfactory state. Of course, disc materials *could* be standardized, but they are not. Not even *test* records are standardized for this purpose. Even if all test records were pressed in a uniform material so they gave identical results with different pickups, we would still be faced with the fact that discs carrying recordings of program material are not necessarily pressed of material identical to the test recording and, therefore, the frequency response of the pickup will not necessarily conform to the results shown under test conditions.

What all of this says is that, while different materials are used for making disc pressings, different pickups will show contradictory differences between the quality of various recordings, as regards frequency response. Comparing disc no. 1 with disc no. 2 on pickup A, may show that disc no. 1 has a high-frequency peak while disc no. 2 gives a very flat frequency response. On the other hand, using pickup B may well show just the reverse due to the fact that the respective compliances of the alternate records happen to suit the mechanical characteristics of the alternate pickups. But using each with the opposite kind produces a peak—or it might equally well introduce an excessive high-frequency loss. This is something that is difficult to predict because so many variable factors are involved.

Unfortunately, manufacturers of reasonably high-quality pickups publish characteristics for their pickups, taken with the "standard test recordings" *which happen to give the best-looking response.* This is by no means "cheating." When there is no standard *record material,* who is to say which is the right recording? All of them show the right light patterns. A typical set of curves is shown in Fig. 710. The deviation at the high-frequency end from 10,000 to 15,000 cycles is by no means a matter of a few db—it can be as much as 10 db or even more between individual recording and pickup combinations.

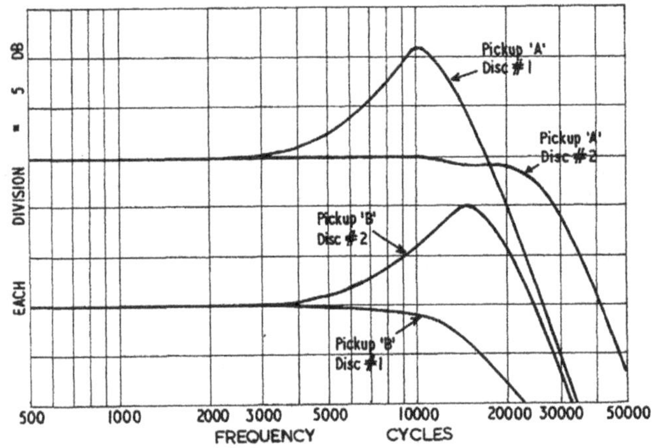

Fig. 710. *A typical set of curves taken with two different pickups on two different test discs, both having accurate characteristics.*

Tape standards

With tape, the situation does seem to be a little better although at the start it may seem to be not so good. There is no convenient means of looking at the response recorded on the tape. We are entirely dependent on exploring the response by means of a playback head. However, careful comparison of results, using different playback heads with varying widths of air gap, and giving different theoretical frequency responses, shows that the results are reasonably consistent with mathematical prediction. This means that we can assume that the recorded characteristic produced on a test tape is within 1, or at the most 2, db of the specified response.

While it is possible, by means of light patterns, to check a disc recorded response even closer than this, the tape does not suffer from the same compliance problem that the disc does. The playback head can be measured on any test tape and the results are much nearer to being consistent than they are with pickups on a test disc, particularly at the high-frequency end.

Methods of equalization

Having obtained standards for measurement, the next step is to provide equalization in audio equipment to meet their requirements.

For disc recording many standard characteristics have been used, all of which differ slightly. This situation arose because each disc manufacturer developed his own standard recording charac-

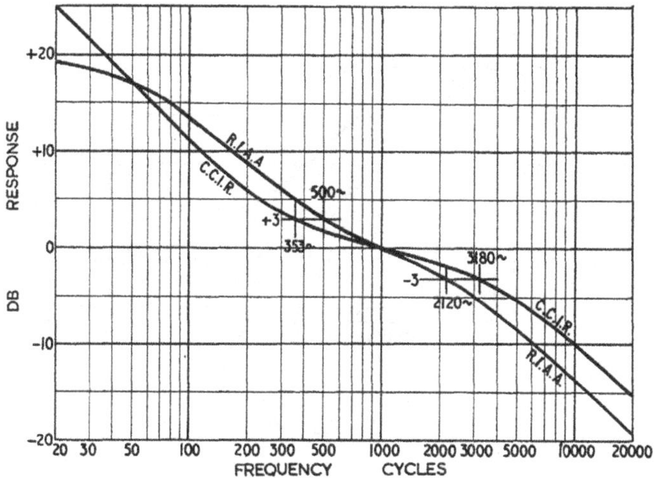

Fig. 711. *The two main standards used for disc recording today. Curves are the required playback characteristics, using an ideal velocity pickup.*

teristic, based on the same fundamental problems but arriving at slightly different variations of a similar conclusion. To reproduce all discs with exactly flat frequency response, a preamplifier needs to have somewhere in the region of 10 equalization characteristics. However, the effort toward standardization is progressing and there are two main standards which come close to fulfilling practically all of the modern requirements in disc equalization.

One of these is that issued by the Recording Industry Association of America (RIAA) and accepted as standard by all of the American record companies. The other standard is known as the CCIR, which stands for the French equivalent of International

Radio Consultative Committee. While almost all American recordings now conform to the RIAA curve, foreign, particularly European recordings, conform to the CCIR characteristic. It is better, of course, to have just two standards than about a dozen. These two characteristics are shown in Fig. 711.

For tape recording two standards (Fig. 712) also exist, one set by the NARTB (National Association of Radio & Television Broadcasters) which is applicable to all American pre-recorded tapes and the other established by the CCIR applicable to most of the foreign recorded tapes.

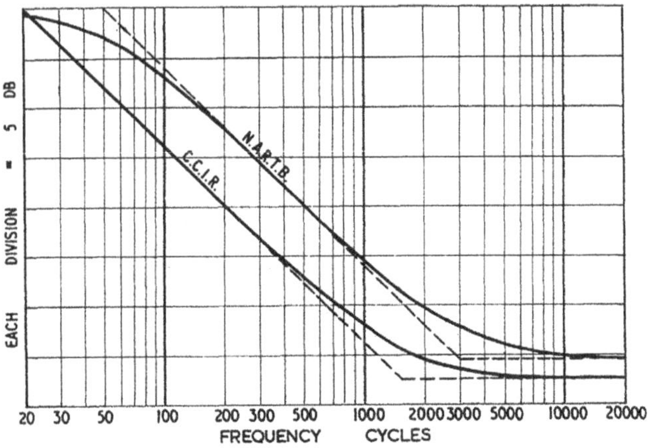

Fig. 712. *The two standards for tape recording (7½ inches per second). Curves are idealized playback characteristics.*

We now come, in considering equalizers, to the part that is more strictly audio—how they are made, the kinds of circuits and their relative merits.

Any equalizer involves a loss in gain. This is because its gain at different frequencies is not uniform, to provide the equalization characteristic. Most of the equalization characteristics provide a low-frequency boost to compensate for the rolloff that occurs during recording. In disc recording, this is to produce an acceptable amplitude of excursion at low frequencies.

In tape recording it is due to the fact that the playback head operates as a "velocity" device (more strictly it is a "change of magnetization" device) while the recording head is a constant-current or, more strictly, a constant-magnetic-amplitude arrangement.

Whichever way you look at it, the result is a system that needs bass boost on playback. This means that frequencies below 1,000 cycles need amplifying more than frequencies at or above that. We are interested in 1,000 cycles because it is the frequency at which the gain of a system is usually measured or specified. Most of the energy in program material occurs in the region from 400 or 600 to 1,000 cycles, hence the apparent loudness is dependent upon the amplification in this frequency range—not on the amplification below 400 cycles or that above, say, 2,000.

However, the fact that, due to the equalization, a greater amount of gain is required at the extreme low frequencies, of say, 50 cycles means that some gain must be sacrificed in the 1,000-cycle region to have the extra gain available at lower frequencies.

Different methods of equalization involve different amounts of loss of gain. Other questions that come into the decision as to which kind of equalizer to use are: the amount of distortion the overall arrangement may introduce into the system; and the effect of the method on the dynamic range available—that is, how much "room" there is in the overall system between the noise level and the maximum level the system will handle. Other important considerations are the accuracy of the equalization characteristic—how closely it adheres to the ideal response required of the system, and also the stability of the accuracy. In other words, how much the frequency response provided by the equalization depends upon tube characteristics or other parameters—whether the equalization is likely to become inaccurate if the gain of a tube changes, for example.

General philosophy

These are the performance considerations. On the practical side comes the question of where in the circuit to put the various "bits" of equalization necessary. The logical thing, to make a system adaptable and foolproof, is to associate each piece of equalization with the particular part of the system for which it is intended (Fig. 713).

For example, equalization for a disc cutter should be as close to the cutter as possible. If it is inserted in a main power amplifier that supplies the cutter, this means that the amplifier is not suitable for any other purpose because the frequency response has been tailored to offset the deficiencies of a particular cutter.

In the playback end it would be ideal to associate the equalization for a pickup directly with the pickup. Certainly the equalization for a pickup should not be used when playing program material from a radio tuner, for example, which should not require any equalization if the tuner is well engineered. This means that the logical place to put the equalization for a pickup is right in the pickup circuit, out in front of the preamplifier, and then to use a preamplifier whose response is absolutely flat.

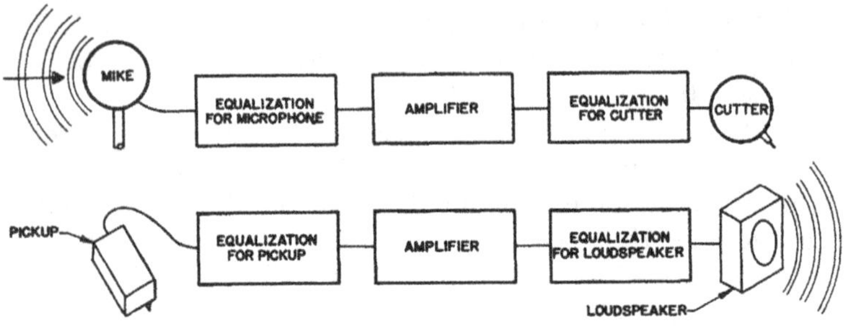

Fig. 713. *Proper equalization associates each correction network directly with the component for which it serves.*

Passive equalizer

Such a unit is called a passive equalizer. The difficulty with this arrangement is that many modern pickups, especially those used for playing microgroove recordings, give such small outputs that the signal–to–noise ratio is very little better than that obtained from a good high-gain preamplifier. Passing the program immediately through an equalizer will cause an attenuation of some 20 db in the signal level, which means that the amplifier will have to pick it up from a point 20 db lower in level. Consequently, the signal–to–noise ratio will be made worse by putting the equalizer directly against the pickup. This is the reason for incorporating equalizers inside preamplifiers, somewhere between stages.

Having decided that the so-called passive equalizer, placed right against the pickup, is not ideal because of the restriction of dynamic range due to noise, the next question is the best kind of equalizer to use in a preamplifier circuit.

Direct type following triode

The most obvious uses a straightforward equalization of a type similar to the passive equalizer but puts it between stages (Fig. 714). The difference in method here depends on whether the stage the equalizer follows is a triode or pentode.

All basic phono equalizers fall into the general pattern of Fig.

715 to provide a bass boost and a high-frequency rolloff. Tape equalizers omit the high-frequency rolloff and may have a high boost instead because of the loss in the playback head. The equalizers, of course, differ in values to suit the individual characteristics required and the circuit into which they are connected.

Fig. 714. *Any interstage equalization circuit is basically a three-terminal network.*

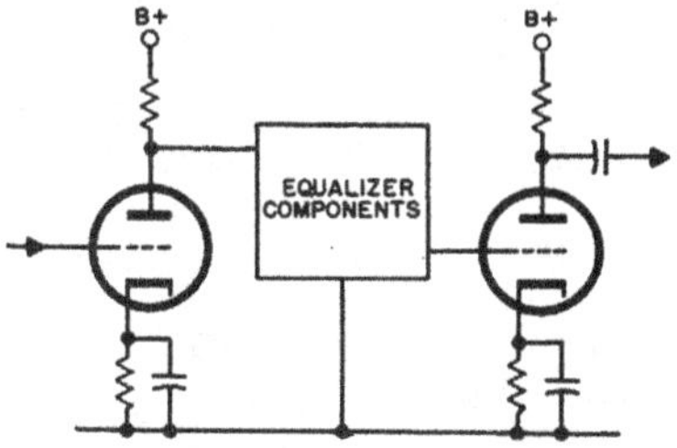

In the interest of providing a minimum loss of gain, where no high-frequency *boost* is needed, R1 can be the plate circuit ac resistance of the preceding stage, instead of inserting an additional resistance for this purpose. In designing such an equalizer, the plate circuit resistance of the preceding stage must not be overlooked. If this is not taken into account, the characteristic achieved will not conform to the specification intended.

Fig. 715. *Basic equalization circuit for phono playback characteristics. R1 and R2 determine the amount of lift in the bass. C1 fixes the point of bass lift in frequency. R_g must be taken into account, although not part of the network. C2 provides the treble rolloff.*

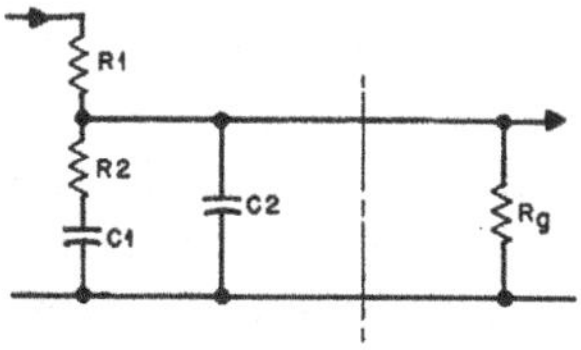

However, if this circuit follows a triode input stage, this means that, at 1,000 cycles (and assuming a 20-db bass boost) the plate of the preceding stage will be operating into a dynamic load about one-tenth of its plate resistance. This almost vertical load line for a triode tube produces excessive second-harmonic distortion. For this reason it is essential to build up the source resistance R1 with an actual resistance value as well as using the plate resistance of the tube, which should be a smaller part of it (Fig. 716).

In this way the operating condition of the triode is protected although more gain is lost than would be without the additional resistance. A triode has the advantage that the inherent distortion is low, provided these precautionary measures are met. However, the fact that the loss is larger than basically necessary for the kind of circuit used means that the attenuation in the equalizer is greater. If the equalizer immediately follows the first stage, the signal level at the second-stage grid will be only a few db greater than that at the input to the preamplifier. This means that the equalization stage can contribute a noticeable proportion of noise to the output.

This deficiency can be overcome by using two stages of amplification before going through the equalizer, but this usually involves additional switching since (in the interest of economy) it will be desirable to use the same stages for radio and other inputs that do not require equalization.

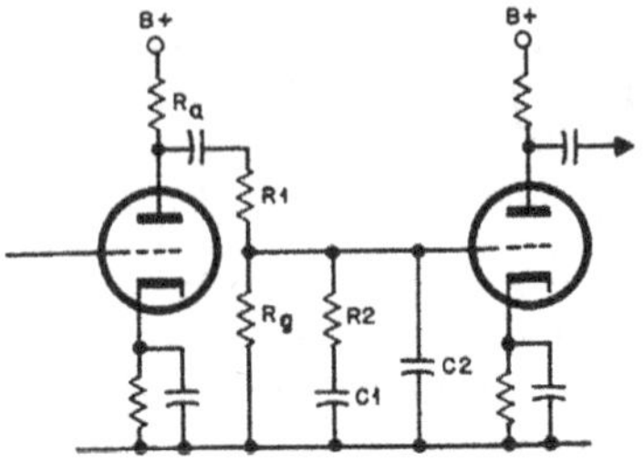

Fig. 716. *The circuit applied between two triode stages. The essential components are numbered to correspond with Fig. 715, except that R1 of Fig. 715 corresponds to R1 in series with R_a, both in parallel with R_g, here.*

When R1 is used, the accuracy of this kind of equalization circuit is unsurpassed by any. Its stability is also good because any changes in tube parameters will not materially affect the equalization characteristic.

Following pentode

If the first stage of a preamplifier is a pentode (Fig. 717), the additional resistor R1 is not necessary in the equalization circuit. The basic source resistance presented by the pentode stage is principally the plate-coupling resistor because the pentode's plate resistance is usually much higher than the coupling resistor.

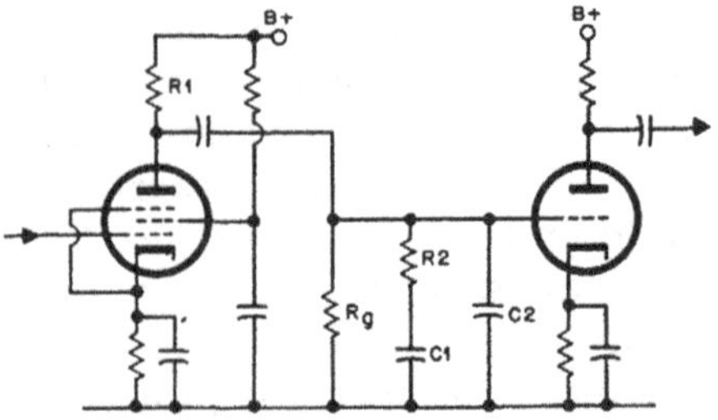

Fig. 717. *The circuit applied following a pentode stage. R1 must be combined in parallel with R_g and the plate resistance of the tube to get the theoretical value of Fig. 715.*

As a pentode stage does not introduce appreciably greater distortion by using a lower value of plate load, the equalization characteristic can be achieved with a value of R2 in the region of one-ninth of the combined parallel resistance of the plate-coupling and grid resistors. This will give the necessary 20-db boost with a minimum of loss.

The extra gain provided by a pentode will improve the signal–to–noise ratio. The accuracy of frequency response and particularly *stability* of the accuracy are perhaps not quite as good as in the triode stage, especially if a high value of plate-coupling resistor is used. Under this condition the plate resistance of the pentode may only be two or three times the value of plate-coupling resistor used and fluctuation with operating conditions can modify the apparent source resistor value by a perceptible degree.

The principal reason for rejecting the pentode stage input, though, is the matter of distortion—the fact that the distortion characteristics of a pentode, even on an input stage, are inherently higher than those of a triode. A secondary reason is that a pentode stage introduces a higher noise level than a triode. This does not invalidate the method of equalization for itself, but because of consideration of the tube type for an input stage.

Separate sections

The foregoing equalization circuits have grouped all of the equalization characteristics into one network. This is not necessary. Under suitable circumstances, the functions can be separated. In fact, for disc equalization, it is sometimes advantageous to put the high-frequency rolloff somewhat earlier in the circuit. It can help out on noise problems by restricting the frequency band in the area where noise is generated.

One very convenient way of doing this can be used if the pickup happens to be inductive. For example, many magnetic pickups have considerable inductance. By loading the input with a suitable resistance value (Fig. 718), the correct high-frequency roll-

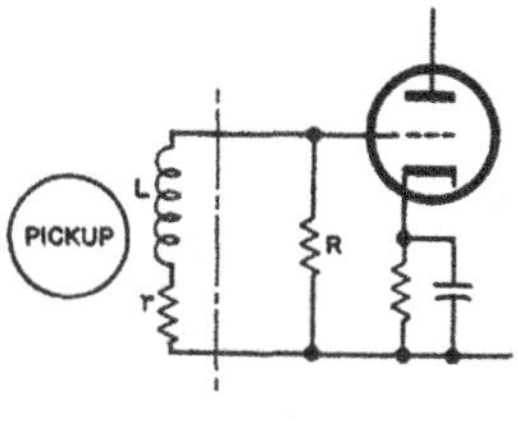

Fig. 718. *Sometimes high-frequency roll-off can be achieved by selecting the input resistor R to combine with the electrical characteristics of the pickup, represented here by r and L.*

off can be achieved right in the input circuit. Then it only remains to use the regular equalization network to produce the low-frequency boost.

So much for the direct type equalization circuits, all of which in general have a similar gain loss. The distortion is dependent upon the operating condition chosen; the dynamic range upon the position in the circuit where the equalizer is inserted. And the stability of accuracy to the required characteristic is good compared to most other circuits, being dependent only upon variation in source plate resistance, where this is of a magnitude that can affect the response of the circuit.

Feedback type

Feedback circuits have achieved a considerable degree of popularity, one reason being an alleged improvement in distortion. We say alleged because, under well-designed conditions, the direct-equalization method can achieve extremely low distortion and even the feedback type is not free from possible forms of distortion. A poorly-designed feedback type will give more distortion than a non-feedback type that has been well-designed. With care in the design, the feedback type can achieve certain advantages in its handling of distortion. It can substantially reduce the more objectionable form of distortion—IM—although it may not be so effective against the lesser evil—harmonic distortion.

Single triode stage

Fig. 719 is a feedback circuit over a single triode stage. The advantage of feeding back over a single stage is that only the feedback network contributes to the response characteristic of the

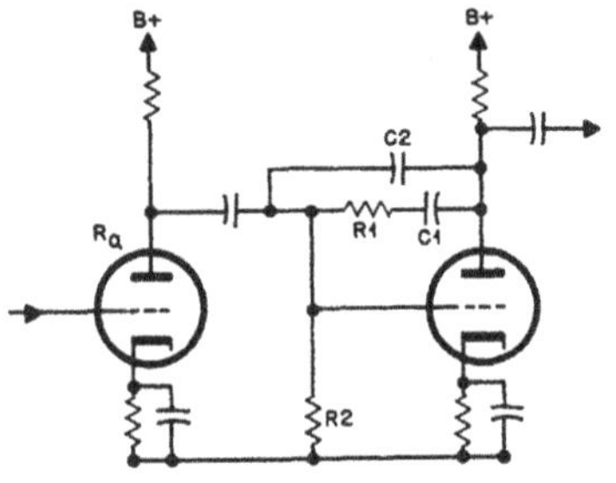

Fig. 719. *A feedback type applied between two triode stages. Design of this circuit runs into serious difficulties. See Fig. 720.*

loop gain around the stage. This means (a) that the characteristic is easier to calculate and (b) that it will achieve greater stability in accuracy for the intended characteristic than a two-stage feedback type.

There is usually some problem with this type in achieving bass boost. This method results in a gain loss equal to the amount of boost required and thus is exactly similar to straightforward equalization. High-frequency rolloff, however, using feedback over a single triode stage, is almost impossible to achieve since 20 db of the gain in the stage has already been sacrificed for low-frequency boost.

The overall loop gain of the stage is not likely to be more than 50, even if a high-gain tube is used. Actually, it is practically impossible to achieve a gain this high, due to the loading of the feedback circuit on the plate of the tube. For this reason, even using a high-slope tube, we are lucky if we achieve a gain of 20 or 30, which is only 6 or 10 db more than the required bass boost. This means we have available only some 6–10 db additional rolloff for the high-frequency end. Rolloff further than this must then be provided in another circuit.

While the loss in the feedback stage itself is the same as in other circuits, it can result in additional loss for the preceding stage. More resistors may be needed to get a satisfactory degree of feedback without losing stage gain. The plate resistance of the preceding stage normally loads the grid circuit resistance and thus requires low values of feedback resistors (Fig. 720).

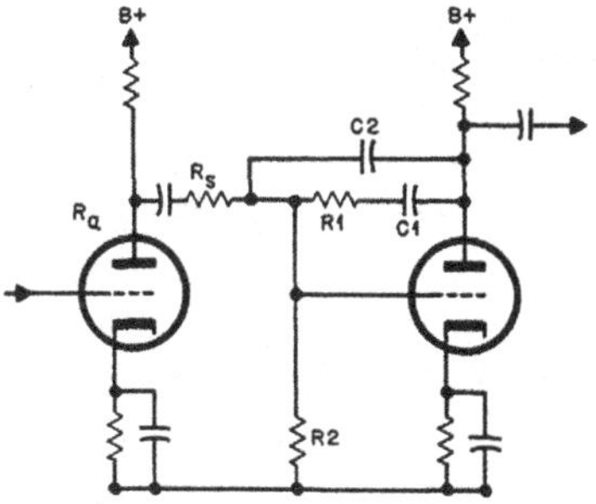

Fig. 720. *To get an adequate amount of gain, and at the same time achieve the necessary range of feedback, the resistor R_s in needed. The effective value of R2 is now the resistor of that number, paralleled by R_s and R_a in series. Without R_s, R_a shunts the value of R2 to a value too low for a practical design. The component numbers in this diagram indicate an approximate correspondence with the same numbers in Fig. 715.*

This circuit suffers from the same deficiences as regards potential signal–to–noise ratio as the placing of the straightforward equalizer after a single triode stage. If a high-gain triode is used, it may be difficult to get a sufficiently high plate-load value to avoid causing some distortion. The feedback may be insufficient to bring the distortion thus produced back to the same order as that achieved in a straight amplifier, using better load values for the tubes.

Single pentode stage

Some of these problems can be overcome by using single-stage feedback over a pentode tube (Fig. 721). The higher gain permits more suitable resistor values for the feedback with less loss of gain. Also the higher loop gain provided by a pentode means that the high-frequency rolloff can be successfully achieved by the capacitor provided for the purpose, without having to resort to additional rolloff elsewhere in the circuit.

However, a pentode, especially operating as a second-stage amplifier, will produce more distortion than a triode stage. While the feedback will reduce the distortion in the 1,000-cycle

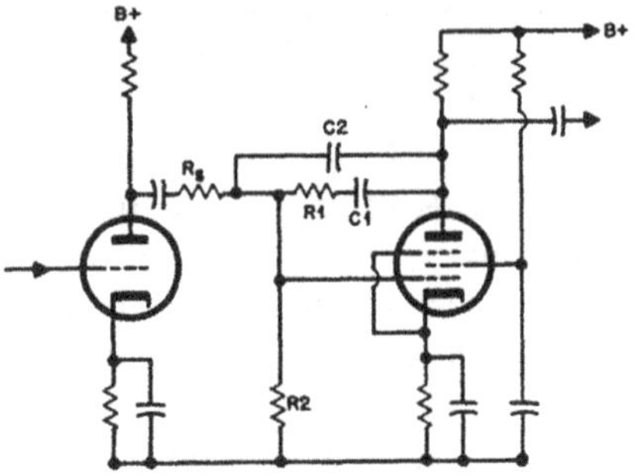

Fig. 721. *Applying the same circuit to a pentode stage gets better results, as far as achieving the desired characteristic is concerned.*

region, *it will not do so at the low-frequency end.* In comparison, the straightforward equalizer, especially following a previous triode stage, will produce less distortion at the low-frequency end and, if anything, only slightly more in the middle of the band.

Thus the pentode feedback stage is liable to produce some distortion in extreme low frequencies. This would normally introduce IM distortion as well as the natural harmonics of the low frequency. However, the feedback arrangement does produce approximately 20-db reduction in the resultant IM distortion. The residual harmonic distortion is probably of little importance since the ear is unable to detect small quantities of harmonic distortion as a separate entity. The principal reason for the importance of harmonic distortion in amplifiers is not the harmonics generated so much as the equivalent IM distortion that usually goes along with it.

Stability of accuracy for the desired characteristic is not quite so good as the type using a single triode stage. The gain of a pentode is more susceptible to change with operating conditions than is a triode and the amount of bass boost is definitely dependent upon the overall loop gain. This, in turn, depends upon the precise gain achieved by the pentode.

Two triode stages

The final type we shall consider here utilizes feedback over two stages of amplification. If these stages are direct-coupled, the same design approach can be used as for a single stage. However, direct coupling involves problems due to the dc-potential differences involved. While it may sometimes be convenient to use direct coupling between a voltage amplifier and a phase inverter, it is not usually convenient to do this between two voltage-amplifier stages, as would be used for providing two-stage feedback equalization.

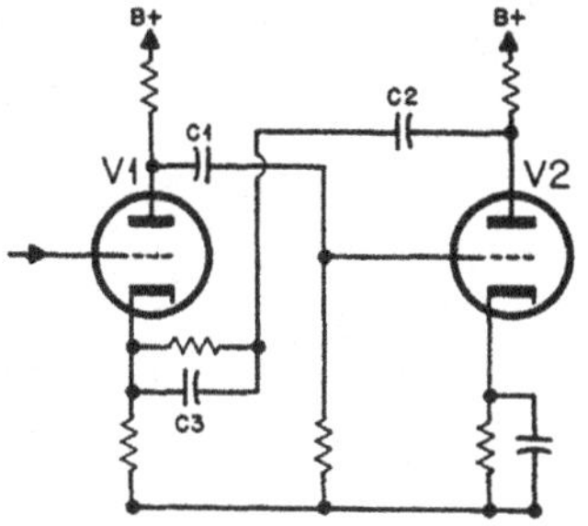

Fig. 722. *The application of feedback over two triode stages. C1 and C2, with their respective associated resistance values, and the overall feedback designed into the circuit determine bass boost, while C3 affects high-frequency rolloff.*

The "forward" coupling capacitor C1 in Fig. 722 has to be taken into account in the design consideration. The method of design consists of considering (1) the loop gain characteristic and (2) the effect that closing the loop will have on the overall *loop* gain characteristic; this will then modify (3) the inverted characteristic of the feedback circuit (Fig. 723).

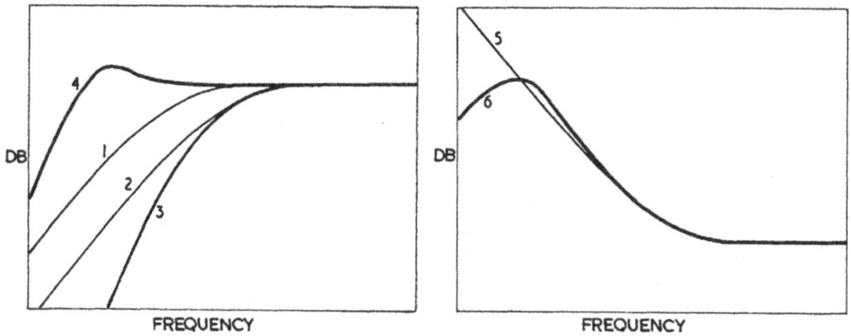

Fig. 723. *Design approach for the circuit of Fig. 722: Curves 1 and 2 show the rolloff achieved by C1 and C2, respectively. Curve 3 is the combined loop gain response, without counting the effect of feedback (measuring the output at the left hand of C2, with a large capacitor bypassing the cathode of V1). Curve 4 is the loop gain response when feedback is added (by removing the bypass). Curve 5 is the response on the right-hand side of C2 compared to that on its left (as flat). Curve 6 is the overall response, due to the fact that the curve at the left of C2 is not flat, but as at Curve 4.*

From here on, there are two approaches. First, the coupling capacitor C1 in the forward part of the amplifier can be so large as to place its effect on frequency response well below the region through which equalization is required. As equalization is usually at least down to 50 cycles, and maybe even lower in some instances, the effective coupling capacitor should maintain

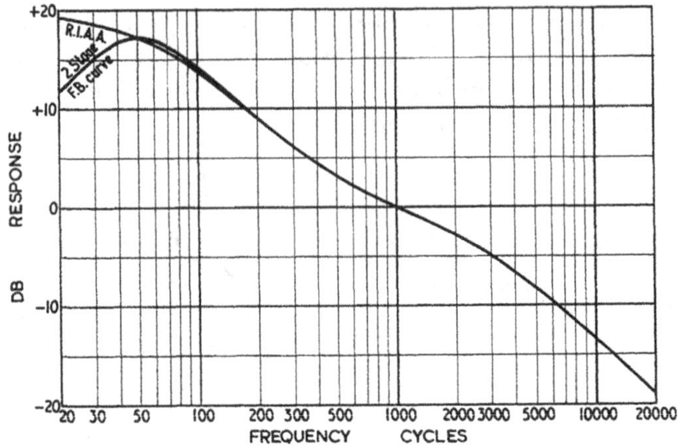

Fig. 724. *Best approach to the RIAA curve obtainable by the methods of Figs. 722 and 723.*

the response down to the region of probably between 1 and 5 cycles. Even then it may invalidate the response by the phase shift it introduces into the overall gain in a way that will not occur where this coupling capacitor can be avoided.

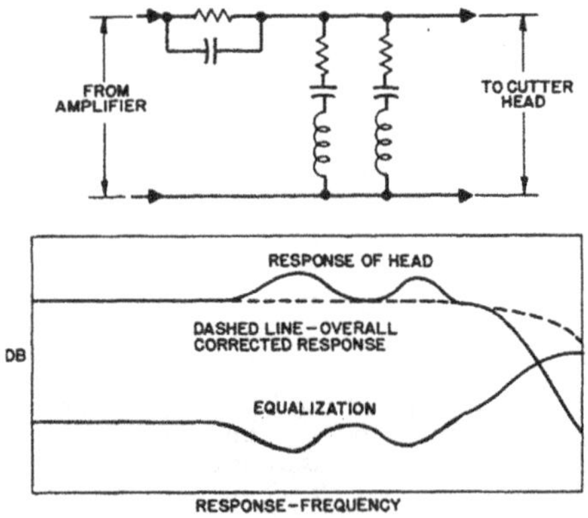

Fig. 725. *A typical equalization circuit to correct the response of a cutter head. The upper part shows the necessary circuit, while the lower part shows the relevant responses, producing a good overall approximation to flat.*

A more convenient method is deliberately to take account of this capacitor in the overall loop gain characteristic so that it contributes to the useful response. This can help to redeem some of the gain loss in the circuit.

For example, normal equalization can be achieved—approximately at any rate—with less than the normal gain loss by this means. However, it does not achieve this result without some modification of the *accuracy* of equalization (Fig. 724).

For some systems this may be advantageous as it produces a rapid rolloff below the lowest frequency to which equalization is carried, in this case shown at 50 cycles. This can be useful in eliminating rumble and other undesirable components from some systems. However, it does somewhat invalidate the accuracy of the equalization characteristic. This must be accepted as a compromise condition.

The distortion of such an arrangement will be similar to that of the pentode stage. The possible signal–to–noise ratio is in the same region also. The principal deficiency of this arrangement is the stability of accuracy. The exact frequency characteristic achieved is considerably dependent upon maintaining the precise loop gain on which its performance is calculated. Variation

Table 7-1. RELATIVE PROPERTIES OF EQUALIZER CIRCUITS

Circuit	Gain Loss * (db.)	Distortion low freq	Distortion mid freq	Distortion high freq	Noise Level	Accuracy	Stability of Accuracy
Passive, following pickup	0	none	none	none	very poor	excellent	excellent
Interstage, following triode	6–10	very good	good	good	good	excellent	very good
Interstage, following pentode	0	fair	good	good	very good	excellent	good
Two-part, rolloff before 1st stage; boost interstage	6–10 or 0	very good or fair	good	good	very good	good	good
Feedback, single-stage triode	6–10	fair	good	very good	fair	good	very good
Feedback, single-stage pentode	3–6	fair	good	very good	very good	very good	good
Feedback, two triode stages	−3 †	good	very good	very good	very good	poor	poor

* Figure is the additional loss of gain due to the type of circuit, taking the loss needed to obtain low-frequency boost as essential.

† Negative because the loss is actually this much less than the other circuits.

in the exact loop gain not only shifts the amount of equalization provided (by altering the amount of feedback) but it also shifts the *frequency* at which the equalization occurs. The variable interaction effect between the parameters of the circuit is responsible for synthesizing the correct response.

The equalizing circuits herein discussed have been confined to "standard" characteristics. Where equalization is required for deficiencies in response of, say, a cutter head, more complicated circuits may be required, tailored to the individual problem. Apart from possible inclusion of resonant components (Fig. 725), the main considerations in using a circuit are very similar to those already discussed. Following is a tabulation of the conclusions about the various types of equalizer circuit discussed.

chapter 8

speaker distribution and crossovers

ASSUMING a satisfactory preamplifier, amplifier and various elements for the reproducing system, the main problem in achieving *realism* centers in the speakers. Original sounds come from all kinds of sources which may be constructed of metal, wood, string, gut or other materials. They may also be small or large. Each sound source has its own "character," and yet we expect a speaker system to reproduce different sounds with *uniform fidelity*.

In the various stages of amplifiers, the principal requirements are a uniform frequency response overall (at any rate when the equalization has been inserted) and freedom from distortion. But when such an almost perfect system is fed into various speakers divergent results can be noticed.

Kinds of reproducer

For example, a speaker consisting of a diaphragm connected acoustically to a number of pipes of different size resembling organ pipes would give an extremely good rendition of organ music because the resonant effect of the pipes would tend to amplify the pipe sound. If the choice of pipe dimensions and positioning were good, such a speaker might well have an extremely flat frequency response. And yet, imagine the spoken voice over such a speaker! It would sound like a man talking inside the airduct of an organ while you were listening outside.

Another illustration: trumpets or horns tend to project the higher frequencies in a narrow path, while the lower frequencies are spread out; the horn type of tweeter unit has a similar directional characteristic so a multiple system using a horn type tweeter can be expected to give realistic reproduction of trumpet or horn music. But it may not be so realistic when it endeavors to reproduce string music, the original sound source of which radiates from the surfaces of stringed instruments. For this purpose a speaker using a plastic diaphragm might give the most realistic reproduction, but its presentation of horn or trumpet music might be quite lacking.

An electrostatic unit also gives a characteristic tone. Because its diaphragm is metallic, it should sound particularly realistic when reproducing sounds like a cymbal clash.

It is impossible, in an orchestral recording, to separate the original components of the music coming from strings and horn type instruments and feed them to different types of speaker units, because the composite frequencies making up the tones from the different instruments overlap. For this reason we have to settle for some compromise that reproduces uniformly and realistically the sounds we want to hear.

The faculty of hearing helps

Fortunately we can be helped considerably by certain properties in the human faculty of hearing. While this faculty is extremely critical in telling the difference, for example, between the kind of radiation from a trumpet and that from a violin, it can also be extremely tolerant. This can be illustrated by considering what happens when a poor public-address system is used in an auditorium.

You may notice at first that the instruments—and even the human voice—reproduced over the system tend to exaggerate a few resonant frequencies. At first these may seem quite irritating, almost to the point of making the sound unintelligible. But, by the time the program is over, you will probably have become so used to the resonances that you find little difficulty in understanding what is said.

Another illustration, a little nearer the realm of audio, is the development of sound reproduction itself. In the early days of acoustic phonographs, reproduced sounds were considered "almost lifelike." Then came electrical reproduction. This gave a "new dimension" to sound reproduction and once again it seemed extremely lifelike, compared to what we had become accustomed. This experience has been repeated many times between then and now, and still we are looking for greater realism.

This highlights the fact that the human faculty of hearing is at once tolerant and critical: tolerant because we can accept the sounds we listen to as realistic and critical because at the same time, we are looking for further improvement.

Big or small?

Original sounds are radiated by sources of different sizes. This is something that is difficult for one speaker to simulate uniformly well. A full orchestra has sound coming from all over the orchestra. If you listen to a program in an auditorium you have an impression of spaciousness in the sound source. On reproduced orchestra music, if a small speaker is used, all the sound comes from a relatively small round diaphragm, say 8 or 10 inches in diameter. So the sound you hear gives the impression, at best, of an orchestra as heard through an 8- or 10-inch hole in the wall.

At the same time when you listen to someone speaking in a small room or studio, the sound originates from an area much smaller than the same speaker.

While a multiple-unit speaker, with the woofer in one place, the squawker or mid-range unit in a different position and a tweeter in yet another, can provide an effectively large source for reproducing orchestral music, it still is a little unrealistic. The distribution of the component parts of the sound "pattern" is still not that obtained with the original orchestra. In the original orchestra, the different sources of sound contain overlapping frequencies: the string section of the orchestra, which may be located to the left, contains the same band of frequencies emitted by the wind section, which may be at the right.

The speaker, however, which does present a large source area, brings the lower frequencies from the woofer, the middle frequencies from the middle range unit and the higher frequencies from the tweeter, *regardless of whether these originate with the string or wind section of the orchestra.* So the distribution pattern is quite different.

However, it is possible to get a better sense of realism with a multiple-unit speaker (or even with more than one single-unit) than with a single relatively small speaker.

Intermodulation

The original reason for introducing multi-unit speakers—one for the low and one for the high frequencies, and later division into more than two sections, to include a mid-range unit—was not primarily connected with the matter of apparent size of source. It was for another reason—the fact that the speakers themselves introduce more distortion than most amplifiers do. This distortion is evident by the intermodulation it causes.

Reproducing just two frequencies at the same time with a nonlinear device will generate spurious frequencies. Program material has many more than two frequencies, so any unit in a system that possesses this form of distortion will cause many frequencies not present in the original program to appear. The most objectionable form of this occurs when low frequencies and higher frequencies are reproduced over the same channel.

A speaker diaphragm moves a considerable distance at low frequencies, hence its sensitivity to higher frequencies may change during the low frequency's excursion. This could be due to nonlinearity of the compliance of the spider, the surround material, or nonuniformity of the magnetic field in the airgap through which the moving coil travels. The net result is that the large-amplitude low frequency modulates the high-frequency tones being reproduced at the same time.

The effect is one that may often have been noticed on ordinary reproduction: the higher frequency tones go dithery whenever there is a loud low note. So, separating the frequencies and feeding them to different speakers avoids some of this problem by presenting the low-frequency high-amplitude tones over one unit and the higher-frequency tones that might get modulated over another unit that does not receive the low frequency tone.

Integration

The next problem is one of producing an overall sound that is well integrated, particularly for the reproduction of spoken program material, with a male voice in particular. It does not sound natural if the deeper, throatier components of the voice come from the woofer, the main body of speech from mid-range unit, while the sibilant sounds get squirted from the tweeter in yet another direction. The three sources of sound should be blended together to give the impression of a single unit source. The coaxial and triaxial type units are the better forms of multi-unit speaker for achieving this objective.

More than one speaker

Because of the conflicting requirements for one speaker, some prefer to use more than one, selecting different types. In this way, by suitable adjustment of balance between the units, each kind of program material can be reproduced with a compromise realism so the overall effect is better.

If the levels are reasonably well matched, the tolerant capacity of the ear helps considerably by psychologically associating voice production and solo program material with the well-integrated speaker and the wide-spread sources of sound with the multi-unit variety. This, of course, can happen only if the levels are right so the ear can hear both units at once.

Because speakers differ in sensitivities, it will often not be sufficient just to connect the two units in series or in parallel, although they may have the same impedance. Two 16-ohm units, in parallel, can be applied to the 8-ohm output of an amplifier. But this may result in one sounding considerably louder than the other, although both receive the same amount of power. This is because one unit is more efficient than the other.

To overcome this, the best plan is to insert some resistance in series with the unit which sounds louder. It will require only a few ohms—say 4 or 5—to balance things. Use a 20-ohm wirewound potentiometer connected in series with the speaker which reproduces the loudest (Fig. 801) and adjust for the most pleasant balance of reproduction of various kinds of program material, as judged sitting in the room at an approximately equal distance from both speakers.

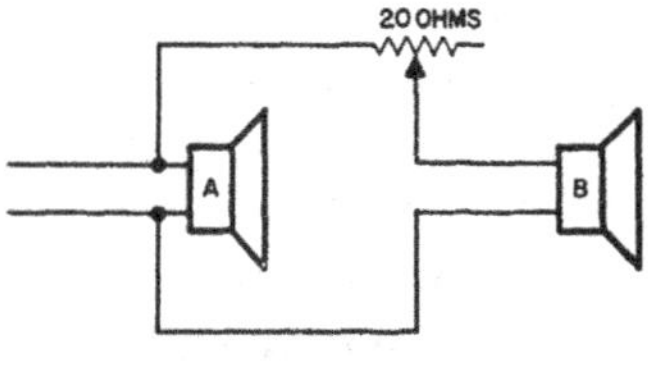

Fig. 801. *If speaker B is more sensitive than A, the two can be matched by a series resistance.*

The effect of series resistance will deteriorate the damping of that speaker which is of special value in the low frequency resonance region (40—130 cycles, according to individual type). A resistance of 4 ohms in series with a 16 ohm speaker limits the damping factor to 4, even if the amplifier has infinite damping. If the speaker needs better damping than this, dual operation is not possible without a compromise. However, the effect of the extra speaker will probably offset the loss in performance in the one unit.

Try reversing the connections of one units and select the connection which gives the best realism—not a sense of "sound in a vacuum." This will be correct phasing.

This works out nicely if both units use an impedance of 16 ohms, or maybe even higher, so they will match conventional amplifier output impedances by being connected in parallel. Sometimes, however, the speaker units may not use convenient impedances. For example, if the speaker impedances are each 2 ohms and the amplifier matches to 4 ohms and higher, it is obviously necessary to connect the units in series to provide a satisfactory match.

From an ideal standpoint, this is not as good as the parallel method. When speakers are connected in parallel across an amplifier output with a high damping factor, each is being fed from a low source resistance or a virtually constant voltage. On the other hand, when speakers are connected in series to an amplifier with a high damping factor, the combination of two units gets the constant voltage. Because of the individual impedance characteristics of the units the frequency response delivered to the respective units will not be the same as it was with parallel connections.

Series operation also means that each unit can, at best, be operated with a damping factor of 1, and the source resistance it sees is far from a resistance. So even more queer things can happen to damping than by inserting a resistance in series with one speaker. But, again, the advantage of having two units may still outweigh this detrimental effect in overall performance.

Occasionally the series method may sound more pleasing than the parallel connection, due to some coincidence in the relationship between their respective frequency and impedance responses. But it is more likely to result in inferior reproduction although in practice that may not be readily noticeable. If the series method is the only convenient method of connection, certainly don't refrain from trying it merely because in theory it is not the right one. Results may be quite acceptable and will almost certainly be superior to the use of only one speaker.

If the speakers differ in impedance and sensitivity try one of the arrangements shown in Fig. 802 for a satisfactory balance. Select the arrangement that comes closest to giving correct balance, then use a shunt or series resistance with the most sensitive unit to get the final best balance. (See Fig. 803).

For stereophonic systems it is desirable to use identical speaker systems on each channel. When different speaker systems are connected to the same channel best results are often obtained by using different kinds of unit. Thus the exact response characteristic from the two is slightly different. While this is not a fully correct method of reproduction because it cannot hope to simulate the original sound, it can give a much more satisfying illusion than a single speaker—one that is often not too much inferior to a full stereophonic presentation.

Multiple-unit speakers

When considering a single-speaker system that uses two or more units, an important consideration is the crossover frequency at which transition is made from one unit to another.

The woofer will have to handle all the bass or lower frequencies, right down to the bottom of the audio spectrum. How high, then, should we go in frequency before we change over to a mid-

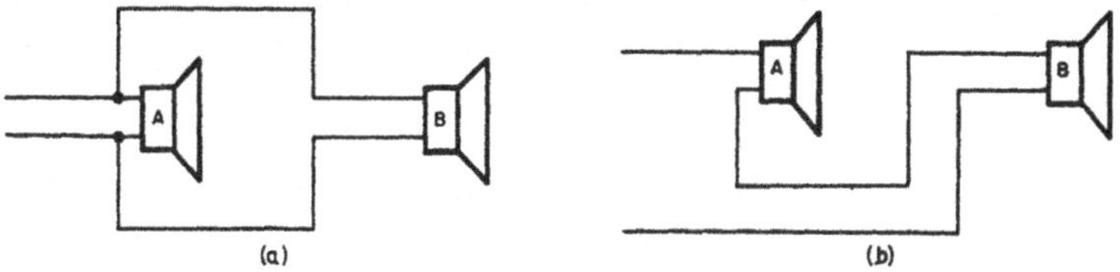

Fig. 802. *If two speakers differ in voice-coil impedance, connect them in parallel (a) or series (b) for the nearest to correct balance in volume level.*

range unit? The principal thing here is to choose a frequency such that the excursion required of the mid-range unit is never large enough for it to cause serious intermodulation distortion.

Where the mid-range unit is of the cone type, the frequency is usually not very critical and any convenient value between 400 and 1,000 cycles may suit.

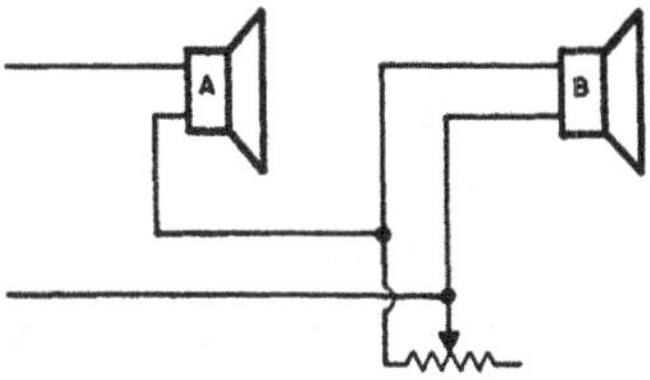

Fig. 803. *If the series arrangement of Fig. 802-b is preferable, the job can be finished by an adjustable resistor in parallel with the more sensitive unit B. If the parallel connection is best. use a series resistor (see Fig. 801).*

Where the mid-range unit uses a horn type speaker—commonly called a squawker—the frequency is a little more critical. Any attempt to produce frequencies below the cutoff designed for the horn will result in excessive excursion of the voice coil in the drive unit, causing distortion. So the crossover frequency should be above the low-frequency cutoff for the type of horn used. This is usually stated in the manufacturer's literature.

While some use a small-diameter cone for the tweeter, the much more common practice is to use either a horn or an electrostatic unit for the extreme high end. Where the cone type is used, the crossover frequency is not very critical. But for the horn type unit we again have the question of the cutoff frequency.

Where an electrostatic unit is used, the matching arrangements are a little more critical. The electrostatic unit requires a high polarizing voltage; and also a high-impedance feed from the amplifier rather than the low-impedance feed used for the other unit. The instructions provided with the electrostatic tweeter are the best to follow for a good match.

Modern high quality electrostatic units come complete with a built-in matching transformer whose primary has a nominal impedance similar to most amplifier outputs. But this does not alter the fundamental situation. A transformer cannot transform a capacitance into a resistive impedance. It merely transforms the capacitance to a different value, suitable for applying across the low impedance circuit. In this case the nominal impedance is chosen by the speaker unit manufacturer so that, with an amplifier of average damping factor, the high frequency response is satisfactory, and the sensitivity comparable with that of an average dynamic unit.

These are conflicting requirements for the electrostatic unit to meet. At the nominal impedance for the JansZen unit, the amplifier must have a damping factor of at least 2 to avoid high frequency rolloff. This could be achieved by connecting the electrostatic matching transformer to a tapping on the amplifier lower than its nominal rating, if the damping factor on the rated tapping is too low, but this will degrade the sensitivity of the tweeter compared to the other units.

Peaky highs

An important point to watch for is the possibility of high-frequency peaking introduced by the use of an electrostatic unit. This should also be watched for with any other kind of tweeter. This effect can be due to poor tweeter design, but often it arises from the peculiar characteristics of individual amplifiers. Do not confuse this effect with the high rolloff due to inadequate damping factor. Actually both can happen in the same amplifier, which is an extremely undesirable effect. The condition we are now talking about is due to a characteristic of the amplifier, rather than to the method of feeding the speaker in question.

These amplifiers have been designed to give a flat frequency response and satisfactory square-wave or transient reproduction, *when feeding a resistance* load. The usual high-frequency tweeter provides an inductive load at the high-frequency end due to the voice coil inductance, while the electrostatic tweeter provides a definitely capacitive load. In various amplifiers one or the other of these may tend to produce high-frequency peaking, due to the phase shift it introduces in the overall feedback loop. The choice of circuit should be chosen with a view to avoiding this trouble. This is covered in the chapter on feedback. Otherwise the only thing to do is to try different amplifiers to find one that gives satisfactory performance into the chosen loudspeaker combination.

Balance

Not only is it necessary to divide the frequencies correctly, but also to maintain correct balance in the reproduction. A well-designed crossover will ensure that uniform energy is fed to one or other of the speakers throughout the frequency band. But if, for example, the squawker and tweeter, because they are of the horn variety, are more efficient than the woofer, then the arrangement will seem to be bass-weak because the woofer does not deliver as much sound energy for the same input as the other units. To get a flat overall response, the mid-range and tweeter units will require attenuation by the insertion of one of the small L-pads available for this purpose.

A simple series resistance, as suggested for balancing wide-range speakers, is not adequate here because it will cause mismatching to the crossover unit. A speaker of 8 ohms, for example, with a resistance of 4 ohms in series, makes a total resistance of 12 ohms. This would compensate for a 2-db difference in sensitivity—it may be necessary to do even more than this. If the crossover is designed for 8 ohms, then there will be considerable mismatch resulting in a distortion of the frequency response in the region of crossover, apart from the question of obtaining uniform level in the main part of the band. For this reason an L-pad of the correct resistance to go with the particular speakers being used should be inserted. In fact it is good to have two in a three-way system; one for the mid-range unit and one for the tweeter.

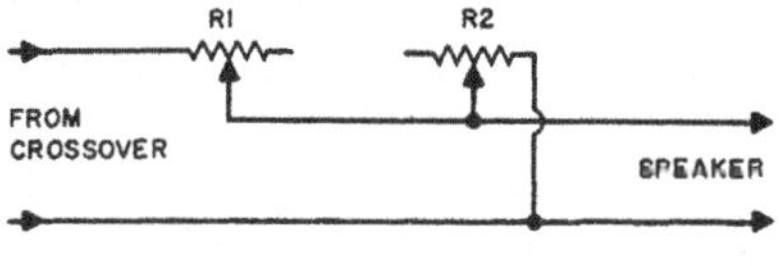

Fig. 804. *A properly designed L-pad maintains the correct resistance to load the crossover by varying R1 and R2 at the same time. The resultant value is always the same as that of the voice coil alone.*

The L-pad maintains matching by using series and shunt resistors at the same time, so the resistance "reflected" to the crossover is always correct (Fig. 804).

In making these checks, phasing is also important. Try connecting the units in alternate ways to make sure you have the connection that gives best realism. Reversal of one unit in a multiple unit system spoils realism.

Kind of crossover

In deciding on the kind of crossover to use, determine how sharp the transition should be from one unit to the other, and how many db per octave rolloff is needed in each unit. This is a matter about which there is some difference of opinion. The answer depends to some extent on the type of unit used.

Where individual speakers have considerable deviation from level, there will also be fluctuating phase angles. Hence it is good to effect as sharp a transition from one unit to the other as possible to avoid an even more erratic response from the combination. But where each unit gives a smooth frequency response in the region of transition, such a sharp crossover is unnecessary. In fact it is better to have a more gradual transition so as to avoid some of the difficulties that can arise with the sharper unit.

There are two ways of designing a crossover network. One uses filter theory; the other is based upon what is termed the constant-resistance technique. The latter is preferred because it delivers constant energy to the combined speakers and at the same time provides a constant resistance load for the amplifier, assuming that the impedances of the individual units are resistive.

In practice, of course, this does not happen because there is always some reactance in the impedance of the speaker. But the use of the constant-resistance type results in less error due to the presence of these reactances than does the filter type crossover.

Fig. 805 gives the relative reactance values at the crossover frequency for constant-resistance networks employing one, two or three reactances in the feed to each speaker. With the aid of a reactance chart, the correct values of inductance and capacitance for each crossover frequency can readily be calculated, having decided on the configuration of crossover to use.

Inductors and capacitors

From the viewpoint of the electrical circuit constants, the next question is what kind of components to use. Should the coils be air-cored, iron-cored, ferrite or toroidal variety? And should the capacitors be paper or will electrolytics do?

There are many factors to consider in making this decision. Many will recommend the use of air-cored inductors and paper capacitors, on the theory that these are the only components that cannot introduce any distortion.

However, air-cored coils, especially for the lower crossover fre-

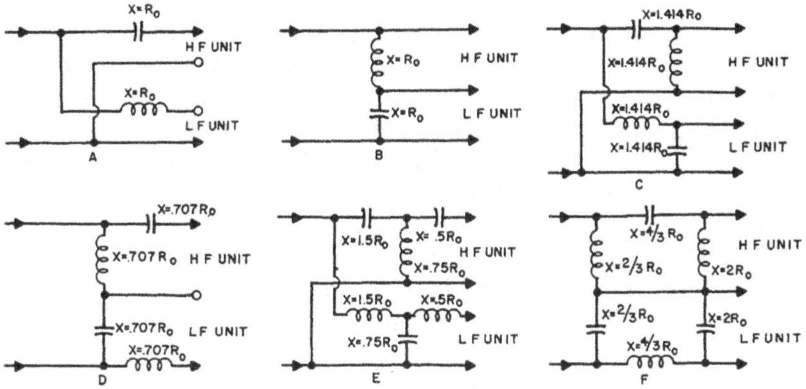

Fig. 805. *Circuits using constant-resistance design values. The reactances indicated are for the crossover frequency, in terms of the working impedance (usually that of the voice coils) of the crossover. Input is from the amplifier.*

quencies, are extremely inefficient. Their Q is low, hence a proportion of the energy that should be fed to the speaker will be expended in the coils. It may also degenerate the efficiency of the crossover as a frequency separating device. The deficiency in energy transfer is probably the more serious one because the low-frequency unit needs to operate at maximum efficiency. If the crossover loses additional efficiency for the low-frequency unit, this means still further attenuation in the other units to obtain a level response and a larger power output from the amplifier to overcome these various losses.

For the low-frequency crossover, an iron-cored inductor, using laminations with an air gap (quite a small one for the lower frequencies), gives the highest efficiency. Its distortion, if correctly designed, is extremely low—lower than that due to other components in a system, as a rule.

The use of some kind of toroid inductor, provided the frequency is not too low, (where toroids become quite inefficient too) is also quite good. These inductors have inherently less distortion than the iron-cored variety although they tend to be somewhat more expensive.

For the capacitor elements, paper tubular or oil-filled types, according to value required, are undoubtedly the best. Once the correct value has been secured, they are not likely to change value with age. But for low impedances the capacitance values are large,

6 or 8 μf being quite common and sometimes even 30 μf or more may be necessary at some points in the circuit. A paper oil-filled capacitor for these values would be physically large and also quite expensive.

It is much easier to obtain electrolytic capacitors in these values. As far as behaving like capacitors, they are quite satisfactory. *If the values are right,* they will perform as successfully in crossover filters as any other capacitor.

The difficulty is that electrolytic capacitors can change their values considerably with time as well as deviating widely from nominal values when first purchased. If a capacitance bridge is available, it is possible, initially at any rate, to obtain the correct values by selection. Pick out capacitors that suit or make them up by paralleling a number of them, until the correct values are obtained.

Electrolytics are originally formed by application of the nominal working voltage until the necessary film is produced on the surface of the active "plate" of the capacitor. The thickness of this film depends upon the composition of the electrolyte and on the forming voltage used.

When electrolytics are stored, the electrolyte film tends to thin down, increasing the capacitance and reducing its working voltage. When such a capacitor is placed into service, it should be reformed by applying the working voltage through a fairly large resistance to limit the polarizing current that flows through it until the film has returned to its normal thickness. Then it can be placed into the circuit with the polarizing voltage applied without fear of a surge current either destroying the capacitor or blowing a fuse in the circuit.

But when these capacitors are used in a crossover network, there is no polarizing voltage. Consequently the film tends to deteriorate anyway. This will result, usually, in an increase in the working capacitance. Electrolytics also tend to dry out, unless they are of the hermetically sealed variety, and the drying out tends to have an opposite effect, reducing the overall capacitance.

This is the serious reason for using better types of capacitor, if at all possible. When various kinds of crossovers are considered, the reasons become a little more conflicting. For the simple types using only one capacitor the value is not so critical and thus an electrolytic might be acceptable, but for the more complicated crossovers using two or three capacitors the values are much more critical. Here is where it would be a bigger saving to use electrolytics.

The best solution is either to operate the crossover at an impedance such that paper capacitors can more conveniently be used, with matching transformers following the crossover, or else to settle for a simpler crossover. Fig. 806 shows the arrangement using matching transformers and designing the crossover at line impedance. It is probable that the saving here on the cost of the capacitors, if paper capacitors were used in either case, more than pays for the extra cost of the matching transformers.

Matching transformers of this type are usually of high-quality design, so the distortion will be negligible, especially feeding from an amplifier with high damping factor. The insertion loss will be less than 1 db (more than 80% power efficiency), and will be the same for each unit.

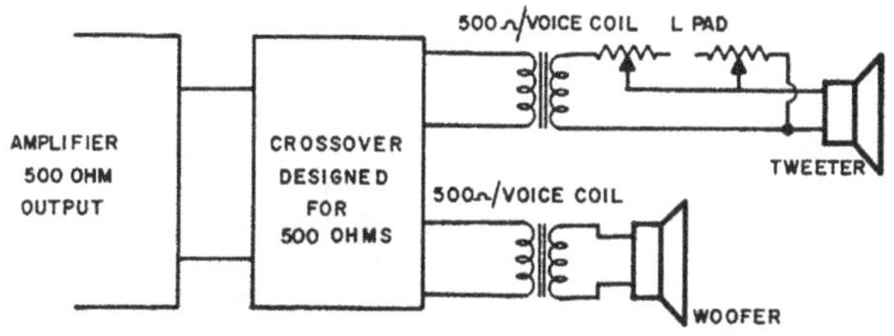

Fig. 806. *For the more complicated crossovers, operate the crossover at line impedance with matching transformers to the individual units. A two-way system is shown here, but the same method can be used for three-way.*

Channel separators

Another kind of separation used for feeding a multi-unit speaker system employs what are called channel separators ahead of the power amplifiers; that is, immediately following the preamplification. One reason for this is that some intermodulation can occur in the amplifiers. Usually, however, the magnitude of intermodulation in a well-designed amplier is much below any other source of intermodulation and, in fact, well below that produced in the average ear!

At the same time, however, some amplifiers with very good intermodulation specifications and that show up well on test give serious intermodulation on some transient signals. This reinforces the reasons for using channel separation before the power amplifiers, although the magnitude of the effect observed is usually quite different from that present in theory.

Another theoretical advantage is sometimes claimed for separate amplifiers. This is that three 15-watt amplifiers can handle a bigger equivalent composite signal than their total rating of 45 watts. This can be derived from a consideration of the theoretical peak power performance of the amplifiers

The peak power of a single sinusoidal wave with an rms value of 15 watts would be 30 watts. If we have three separate frequencies, one in each band, so that one gets amplified by each amplifier (each of which has an rms value of 15 watts) each will have a peak amplitude corresponding to 30 watts in its own amplifier. The total peak amplitude of the voltage amplified will be three times the amplitude of any one. Trebling the amplitude is equivalent to multiplying the power by 9, so the *effective* peak power presented to the loudspeaker system is not 3 times 30, or 90 watts, but 9 times 30, or 270 watts.

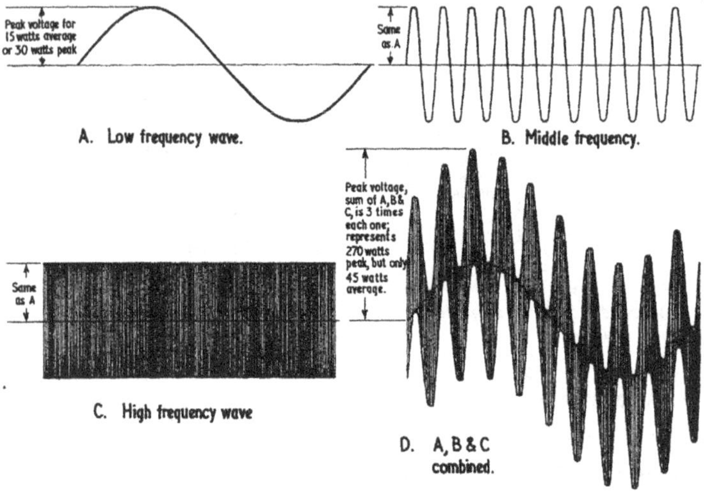

Fig. 807. *Some "theoretical" waveforms associated with use of channel separators.*

To handle this peak power a single amplifier would need an rms rating of 135 watts, instead of the supposed total of 45 watts. Thus the argument goes that three separate channel amplifiers, rated at 15 watts each, can handle as much power as a single 135-watt amplifier (Fig. 807).

This reads well in theory, but not many practical programs contain one sinusoidal tone in each of the frequency bands, each of equal amplitude. More often there will be two or three tones of equal power *in the same band* and only relatively small powers elsewhere, particularly in the range delivered to the tweeter. So this argument losses much of its validity when we think about practical program signals. Most of the power should be provided to handle the low and middle frequencies.

Another argument for separate channel amplifiers is that it

avoids allowing the speaker impedance characteristics to interfere with the crossover accuracy by using only R and C elements in the separator and feeding the speakers from well-controlled individual amplifiers. This is true, but it is also possible to include the speaker reactance elements in the crossover elements of Fig. 805 and thus avoid this disturbance, where it proves critical.

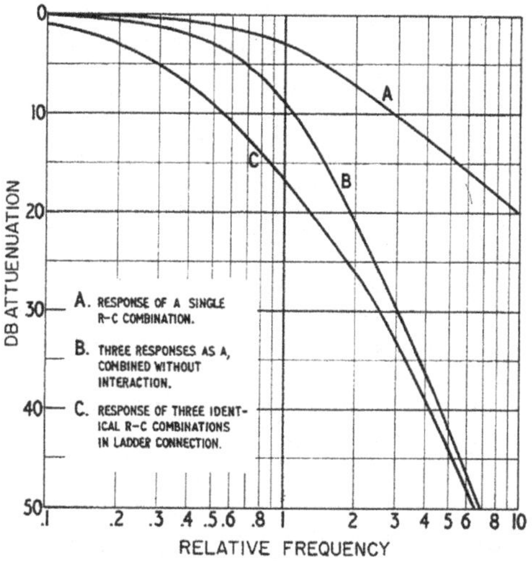

Fig. 808. *Effect of ladder connection of R-C components. A is the response of a single R-C combination. B is the addition of three responses as at A, which would be produced by three responses A in cascade. C is the response produced by three identical R-C combinations in ladder connection. It differs from B due to interaction between components.*

Although it would seem that the best approach would be to get a better power amplifier and not bother with the separation before it, this is still a matter of individual preference because of the number of factors involved.

Because of the relatively low levels involved, about 1 volt at high impedance, it is not practical to use the same type of crossover used for the speakers, because inductances will prove susceptible to picking up hum. They can be shielded, of course, but this will make the unit quite expensive.

As most people who prefer this type of circuit are purists who desire to eliminate any *possible* cause of distortion, this is another reason for avoiding inductances—although they would not necessarily introduce an *audible* component of distortion, if well designed.

This leaves us with a variety of resistance–capacitance circuits. These fall into two groups: those with feedback and those without.

R-C circuits without feedback produce an additive effect similar to a number of separate rolloffs combined. The rolloffs can be spread so that the effect does not occur at the same frequency, but the maximum effect can be achieved by using a number of resistance–capacitance combinations that produce a rolloff at the same frequency.

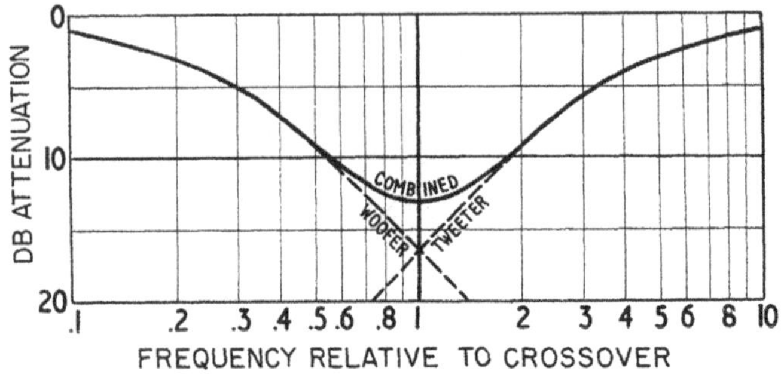

Fig. 809. *Combining high-pass and low-pass three-stage ladders, using identical R's and C's throughout, produces a "hole" in the combined response.*

If the R-C values of a ladder network are identical, then the effect of the successive components interact to "spread" the rolloff characteristic anyway. As a result the mid-point of the characteristic, where the phase shift is half the ultimate, will be accompanied by more than the usual combined attenuation (Fig. 808).

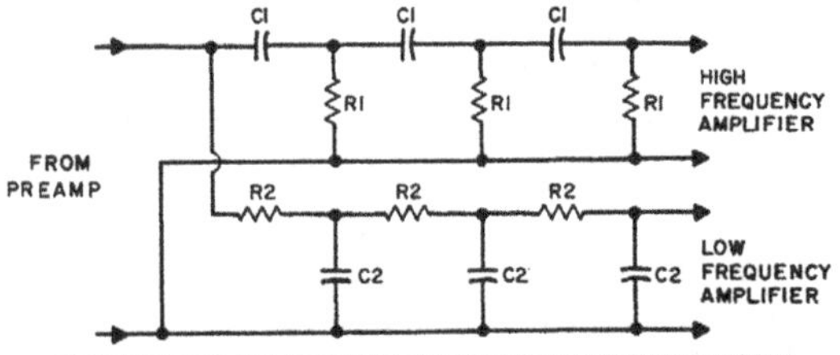

NUMBER OF STAGES IN EACH LADDER		1	2	3	4
VALUE OF X_c AT CROSSOVER	HIGH FREQUENCY FEED: C1	R1	.374 R1	.196 R1	.118 R1
	LOW FREQUENCY FEED: C2	R2	2.68 R2	5.11 R2	8.47 R2
PHASE DIFFERENCE BETWEEN OUTPUTS AT CROSSOVER		90°	105°	110.7°	112°

Fig. 810. *Design data for ladder networks to give approximately constant power.*

For example, a network using three resistors and three capacitors produces an ultimate phase shift of 270° and a midway phase shift of 135°. The loss at the midway frequency due to separate

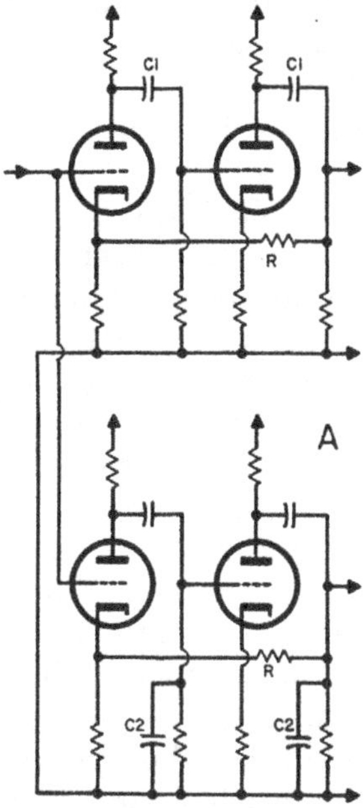

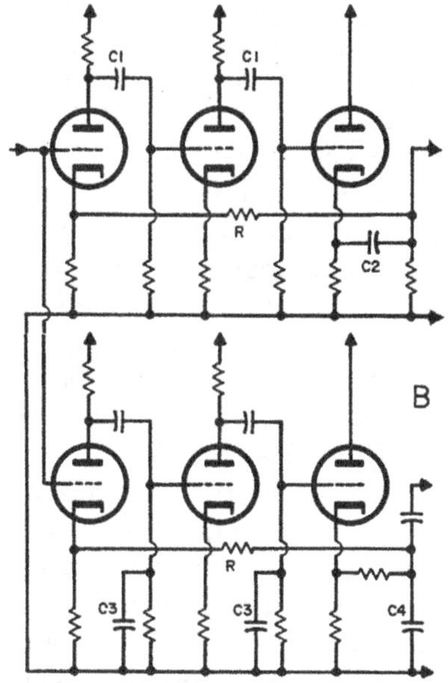

CRITICAL REACTANCE VALUES AT CROSSOVER FREQUENCY (TIMES RESPECTIVE CIRCUIT RESISTANCE)	CI	C2
	1.414	.707
RESISTANCE TO GIVE FEEDBACK OF	R	
	6DB	

A

CI	C2	C3	C4
1.322	.87	.75	1.15
R			
3.67DB			

B

CI	C2	C3	C4
2.614	1.082	.3826	.924
RI		R2	
16.7DB		1.38DB	

C

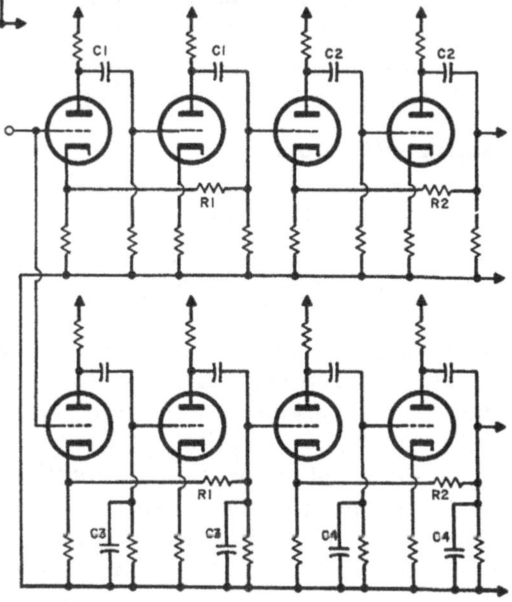

Fig. 811. *Basic circuits for feedback channel separators. Only critical values are shown. Values are given in each case relative to the circuit in which the component is connected. Circuits include no precaution for ensuring minimum distortion, or optimum dynamic range. These must, of course, be taken into account when designing a practical circuit. All that is shown here is intended to illustrate the critical components for obtaining the right frequency response (a) 12 db/octave, (b) 18 db/octave and (c) 24 db/octave.*

noninteracting networks would be 3 times 3 db, or 9 db. But with a corresponding interacting ladder the loss will be 17 db at this point, although the phase shift is still 135°. This is 8 db more than the 9 db that would be provided by combining three separate rolloffs at the same frequency.

This means that making the rolloffs at the same frequency for both high- and low-pass sections of the filter, i.e. using identical R's and C's throughout, will produce a 14-db dip in the response. (Fig. 809). This can be remedied by "sliding" the design frequencies for the two sections together to produce a 3-db loss in each at the crossover frequency. This will give the same total energy at crossover frequency as is delivered to each unit at frequencies remote from crossover.

When this is done, the circuit achieves very close to the constant-resistance requirement, although there is some phase difference in the region of crossover frequency that is not normal to a constant-resistance arrangement. However, the ultimate rolloff slope is not reached until a long way beyond the crossover frequency. The slope at crossover is very approximately 3 db per octave and at one octave beyond crossover it has reached the region of 4.8 db per octave and an attenuation of about 7 db–regardless of how many stages are used (Fig. 810).

Feedback separators

If a sharper slope is required beyond crossover, the only way to achieve this with R-C components is to use a feedback type of crossover. This uses feedback to accentuate the rolloff produced by the individual components. Responses can be achieved that are identical with any of the constant-resistance crossover types, both in phase and attenuation characteristics. Basic circuits and parameters are shown in Fig. 811.

They have the advantage over crossovers in the power output in that they are not susceptible to the reactance loading of the speaker, which reflects only on the output of the amplifier and does not influence the performance of the crossover.

However, to offset this advantage, there is the disadvantage discussed in the previous chapter, relative to equalizers, that they are much more critical of tube parameters in the section of the amplifier over which the feedback is applied. Change in the gain of a tube will change the sharpness of the crossover, the phase shift and attenuation characteristic, by at least as much as loading with a speaker, that contains a reactance, will modify the characteristic of a constant-resistance crossover in the output.

chapter 9

loudness & volume controls

THE question of whether to use a volume or a loudness control is one on which there has been disagreement. There has also been some argument about what is required of a loudness control. Loudness and volume are two names for the same thing. But, when audio was an unimportant part of a radio set, a potentiometer to vary the sound level was called a volume control (which might equally have been called loudness at that stage). Actually, the potentiometer adjusted the *gain* of the audio section, so more properly it should be called a gain control.

But it was noticed as soon as *better* quality reproduction was approached that adjustment of level by a volume (or gain) control changed the apparent *quality* of reproduction as well as just its loudness. The reason was evident in curves published by Fletcher and Munson showing the average dependence of human hearing *sensation* on the intensity of the sound wave causing it (Fig. 901). Obviously, these curves must shed some light on what a loudness control should do, but some difference of opinion has existed as to just what.

Some have concluded that a loudness control needs to compensate for the average hearing characteristic of the human ear. This is a fallacy. Original sounds do not sound distorted because we do not wear a loudness control; they seem normal to us the way they are. So if a reproducing chain, from microphone through recording or transmission, playback or reception to the speaker reproduces all the component frequencies at our ears with an intensity *identical* to the original, it should give the best realism possible from this viewpoint, without loudness compensation.

The unnaturalness of reproduced sound occurs when the intensity of the reproduction is *different* from the original. Suppose

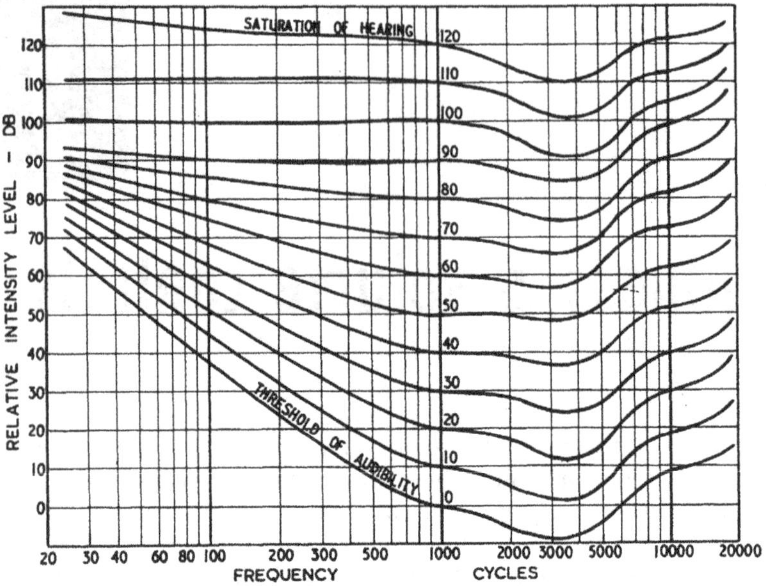

Fig. 901. *Loudness sensation contours obtained by Fletcher and Munson. The shape of these contours forms the basis of the discusion about requirements for a loudness control.*

it is 30 db lower. At 1,000 cycles this will reduce the sensation level by three of the curves in Fig. 901 (called a difference of 30 "*phons*" to avoid confusion with the *intensity* difference). But down at, say, 50 cycles (assuming the original level was on the "70" curve, representing 70 phons, which has a relative intensity level of about 83 db), reduction by the same intensity, 30 db, to 53 db, brings the sensation just above the 0-phon curve, which means it will only just be audible. So the 1,000-cycle tone will be reproduced at a sensation level a little over half the original (40/70), but the low frequency will almost have disappeared. In fact it will probably be drowned out by the middle frequencies being at such a higher sensation level.

So a loudness control, to correct for this change in level, should turn down the 1,000-cycle level by 30 db, but the 50-cycle level by only 11 or 12 db. This will change the apparent loudness at both points from 70 to 40 phons. Some compensation is also necessary at the high end (but not nearly so much) because the curves are almost parallel above 1,000 cycles. So a change in intensity level will produce an almost identical change in loudness sensation over this range.

Recordings are usually made at an intensity level in the region of 70 phons. If a musician plays a pianissimo passage, he drops the different frequencies by a similar apparent loudness because he judges sound by ear (since he is not reading a volume indicator!). So a quiet passage will automatically be compensated by the way the musician plays it. But a maximum level of 70 phons is high for most living rooms. An auditorium or studio has a reverberation characteristic that helps by "averaging out" the sound intensity so it does not seem unbearably loud as it will in our homes. The same loudness in a living room is too much, except for the enthusiast who thinks high fidelity is synonymous with high intensity.

To get a home listening level the music must be reproduced at about 55 phons (although this varies widely with taste!). Or, to use the music as a background for a get-together with friends, a level of about 40 phons will be plenty. However, turning the volume control down by whatever amount is desired makes the reproduction sound thin, due to deficiency in *apparent* bass. Thus the loudness control came to be advocated.

In any system there are two conflicting desirable objectives: versatility and simplicity. The protagonists of simplicity state that, if we have a normal tone control for bass and treble boost (as well as rolloff when desired) the loudness control only gives us another control that does the same thing.

It was argued that, since bass boost is the same as treble cut and vice versa (an over-simplification by modern standards!), only one control is needed, which might called a "balance control." It was also stated that separate equalization is unnecessary because tone controls can take care of this function. Advance in the quality of appreciation has proven these arguments fallacious. The educated ear *can* detect these subtle differences. So any self-respecting system now has an equalization switch or built-in equalization for the most-used characteristic, plus tone controls.

There seem to be two approaches as to whether we need a loudness control and, if so, whether it needs to be additional to the volume control or separate from it.

The first is to use a volume control and adjust the tone controls to suit the loudness at which the program is played. The disadvantage is shown in the curves available in a tone-control system. Very few tone controls will give the *shape* of frequency-compensation curve required to give correct adjustment for loudness. If the control has an adjustment for frequency range and degree of boost at both ends of the response (four tone-control knobs), it will be possible (but quite complicated) to obtain correct adjustment for this compensation. With the simpler control that has one knob for bass and one for treble, the requirements for loudness compensation can conflict with those for other tone-control purposes.

The alternative approach uses a loudness control that produces correct compensation for loudness differences as listening level is changed and leaves the tone control to care for other devations in balance. This approach postulates that, when the reproduced level is the same as the original sound, the loudness control should give level response. To achieve this, we also need a volume (or gain) control to compensate for differences in input level from various program sources—phono, radio or even discs from different companies.

To use this system properly, set the loudness control to a predetermined level, say 50 phons if this is the level at which you normally listen to programs. Then, with the program going into the amplifier, adjust the volume control to your normal listening level. Having set this level, adjust loudness, if desired, by the loudness control. The effect should be just that of changing loudness, without apparent change in quality or balance. This will work quite well and gives pleasant handling, but there is one more question. How do you know when you have the right combination of setting between loudness and volume control?

Strictly speaking, you don't. This is a matter of individual judgment, which may be in error. But the method of control is easier to use, even if not ideal. If part of your loudness-control function is served by the bass control or vice versa (once you have good balance established) the loudness control will give a better loudness adjustment than the other method because loudness differentials are similar over quite a range of loudness levels.

Thus we have two approaches. Of course, a volume control with good tone controls is preferable to a poorly designed loudness control—and some loudness controls do not produce anywhere near the right characteristics. But there is one method that is difficult to justify: a control with a switch that converts it from loudness to volume-control action. If a preamplifier is to be used exclusively with program material in which an input level of 1 mv at 1,000 cycles, for example, always corresponds to a loudness of original program of 70 phons, then no gain control is needed—only a loudness control. If you intend only to replay a program at its original loudness, whatever its source, then you do not need a loudness control, only a volume (or gain) control to take care of differences in electrical level at the input that do not represent

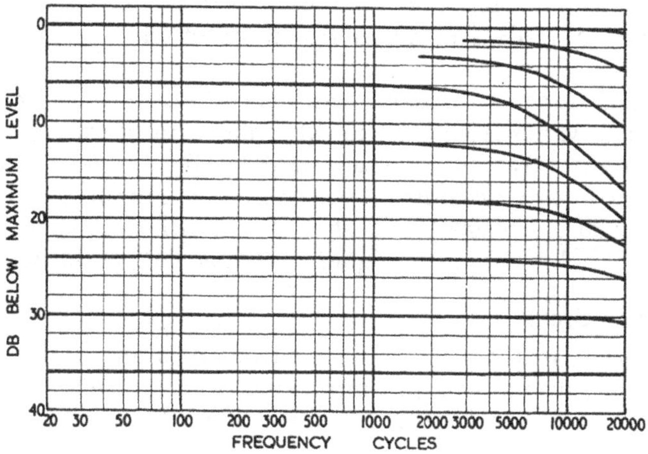

Fig. 902. *The effect of a volume control in producing a high frequency rolloff at some levels. The curves shown represent a 1 megohm control feeding into a grid input capacitance (total effective) of 100 μμf, or equivalent combination.*

corresponding loudness differences. But a preamplifier is not likely to be required for one *or* the other of these requirements. If it is not to be worked under either exclusively—admittedly unlikely—then *both* methods of adjustment should be available simultaneously.

It would seem that the only use for an amplifier with a switched loudness *or* volume control can be this: First, try the program using the loudness control. If the channel level at the input is correct for the gain of the preamplifier so loudness setting will correspond, you are in luck and can use the control as a loudness control. Otherwise, you must switch back to volume control and rely on the tone controls to compensate for loudness differences. Such an arrangement would seem little better than no loudness control.

There is an effect that is occasionally confused with the requirement for a loudness control. Sometimes a volume control drastically reduces high and low frequencies when turned down. The reduction in apparent low frequencies is due to the loudness contour difference effect. But the high-frequency loss is a real electrical loss. It is caused by using too high a value of potentiometer for the volume control, according to the input capacitance of the following stage (Fig. 902).

When a volume control is at maximum, the source resistance of the preceding stage is coupled directly to the grid of the following stage. But as the volume control is turned down, some of the control resistance is interposed between the source and the following grid, causing increased rolloff due to its input capacitance.

The magnitude of the effect will depend on the input capacitance, which varies from tube to tube, and with operating conditions. But the effect can be minimized or eliminated by using a lower value of volume control. According to the input capacitance used, 100,000 to 250,000 ohms should be satisfactory. But in no circuit will a 1-megohm or higher potentiometer be satisfactory.

Position in circuit

This brief review of loudness control operation will help us see what is required in a preamplifier which uses such a control. The position of the control corresponding to, say, the 70-phon

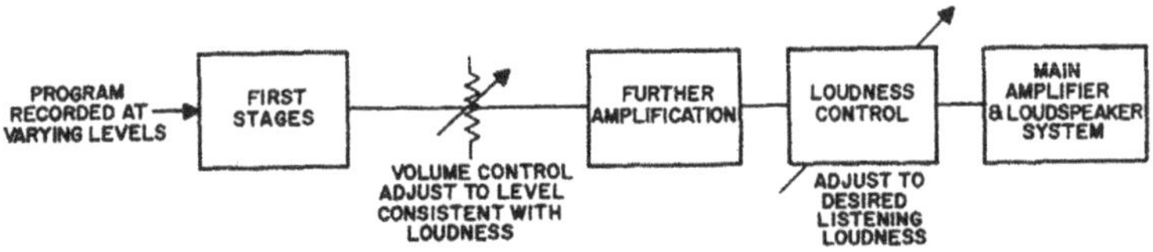

Fig. 903. *The general arrangement of loudness and volume control in an amplifier, showing how to operate the combination properly (see text).*

level, must always result in reproduction at a 70-phon level. Turning the control down 20 phons means that the reproduction should likewise always be at the 50-phon level. Only in this way can the loudness sound right at different settings—or in fact at any setting of the control.

As different recordings are made at varying intensities, even for the same dynamic level in the recording studio, this means some adjustment is necessary to make sure that the reproduction level corresponds to the setting of the loudness control. So the correct operation of loudness and volume control necessitates use of the volume control at some position in the circuit ahead of the loudness control. Fig. 903.

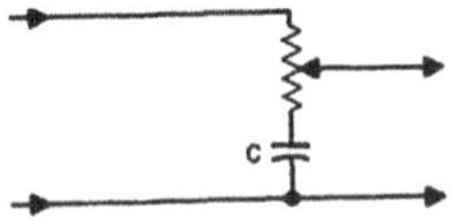

Fig. 904. *The simplest modification to a volume control to convert it into a rudimentary loudness control. The position of the progressive bass emphasis is determined by the value of C, compared with the potentiometer resistance.*

Circuits

Fig. 904 shows the simplest circuit that will produce a variable gain or level adjustment and at the same time alter the bass boost. Fig. 905 shows the family of characteristics for various settings of this control. The obvious deficiency is that it does not have an off position. At the bottom end, it has an indefinite 6-db-per-octave rolloff which gives muffled low-level reproduction.

Fig. 906 shows one circuit which overcomes this by using a

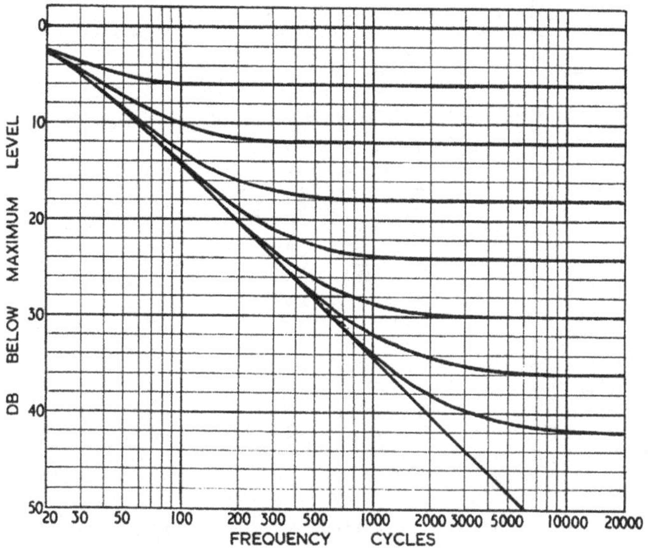

Fig. 905. *Family of curves produced by the circuit of Fig. 904 at different settings. This arrangement does produce the right variation of contour with level.*

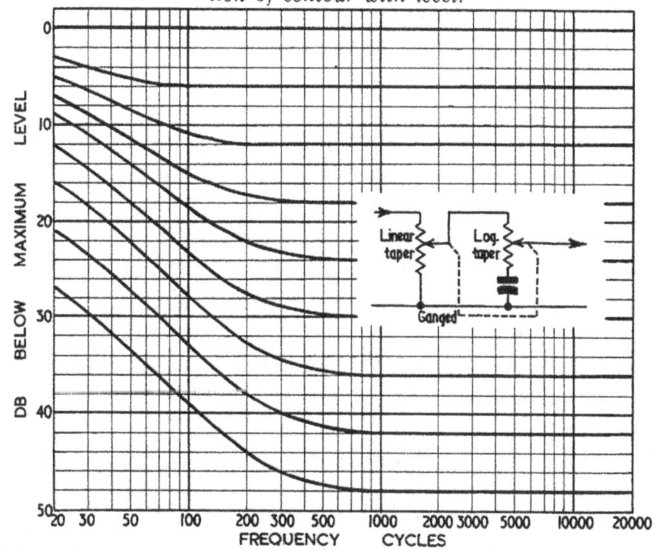

Fig. 906. *A simple loudness control built round a two-gang potentiometer, and the kind of response characteristics this circuit can give.*

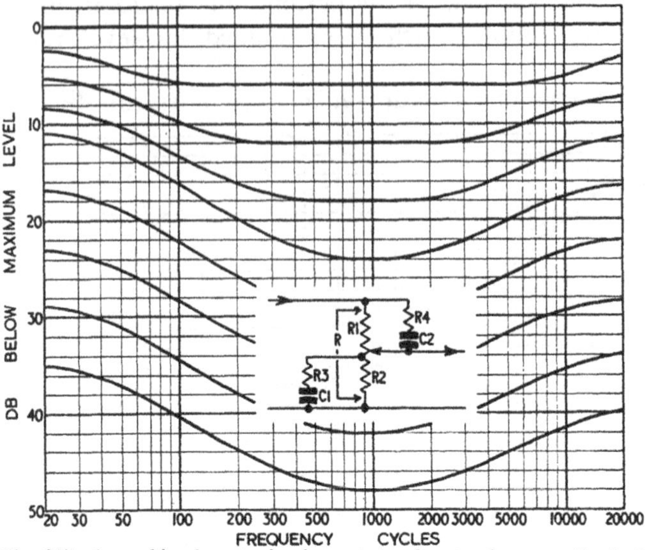

Fig. 907. *A combined control using a tapped potentiometer. Typical values using a 100K potentiometer: R1, 75K; R2, 25K; R3, 3.6K; R4, 47K; C1, .1 µf; C2, .0015 µf. Other values in same proportions would produce similar responses.*

ganged control. The first control, a linear potentiometer, does not have much effect—only 6 db—throughout the upper half of its rotation; most of the gain control in this part of the rotation is due to the second potentiometer, which uses a logarithmic taper. But in the lower part of the rotation the first stage of the control insures that it goes right down to an off position instead of producing a muffled effect.

An alternative method without two separate resistance elements, utilizes a single potentiometer with a fixed tapping. A separate resistor and capacitor are connected to this tapping point to produce the bass-boost effect. This circuit, however, limits considerably the amount of boost achievable. The boost gets progressively greater as the volume is turned down until the slider reaches the tapping point. From here to the bottom of the control the boost remains constant. While this is an improvement on a control with no such compensation, it is far from adequate control for true loudness effect.

So far we have not considered any compensation for the treble end of the response. Unless it is necessary to compensate for high-frequency losses due to the volume-control effect only a small degree of compensation is needed here. But it does definitely improve the presence effect. A practical circuit combining the two kinds of control is shown in Fig. 907. This uses an additional re-

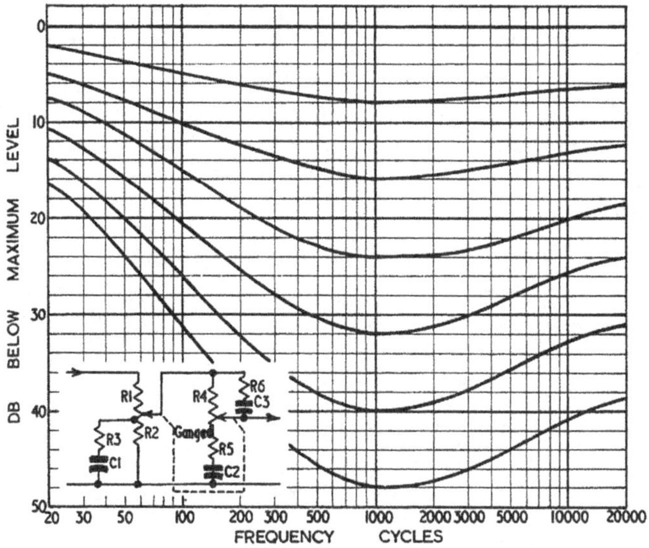

Fig. 908. *A two-gang control giving very good loudness compensation. The curves shown are for rotation over the upper half of the control movement. The lower half will produce additional attenuation without much further compensation. Typical values: R1-R2, 75 + 25 = 100K; R3, 3.6K; R4, 500K; R5, 22K; R6, 220K; C1, .2 μf; C2, .015 μf; C3, 100 μμf.*

sistance and capacitor to produce an ultimate of about 10-db lift to correspond with the low-frequency boosting effect.

However, none of these circuits provides adequate boost at the low end. This should get progressively more down to the low frequencies if it is to produce really high-fidelity low-level effects when turned down. To achieve this we really need a two-stage arrangement (Fig. 908). Here the tapped control gives quite a satisfactory additional boost at the extreme-low-frequency end. The main boost is provided by the untapped potentiometer, padded out with an additional resistor to achieve the correct rate of change in combination with the tapped section. Variable high-frequency response is achieved by using a capacitor and resistance to shunt the upper end of the second section.

Continuously variable or stepped

Many other loudness controls have been developed with varying degrees of accuracy and circuit complexity. There is one more question to consider: whether the loudness control needs to be continuous or whether certain fixed settings would be preferable.

Some loudness controls consist of a twenty-position, 2- or 3-pole switch, like those used for professional attenuators, into which separate networks are connected for each position to give an appropriate compensation, corresponding to the Fletcher-Munson characteristic. This can produce a loudness control of extreme precision but also one that can be very expensive due to the large number of components.

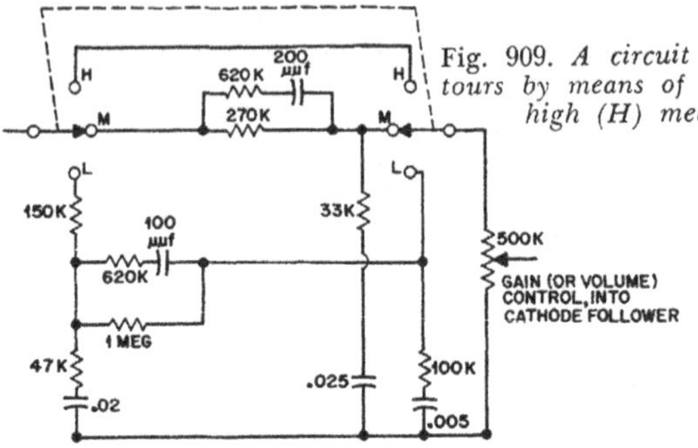

Fig. 909. *A circuit for giving three definite contours by means of a selector switch, identified as high (H) medium (M) and low (L).*

There is a good argument for using a switched control, in which compensation is provided only for three or four levels. On the basis of a normal recording level of about 70 phons, the upper position passes the signal through without any compensation whatever as the top position of a good loudness control does. The

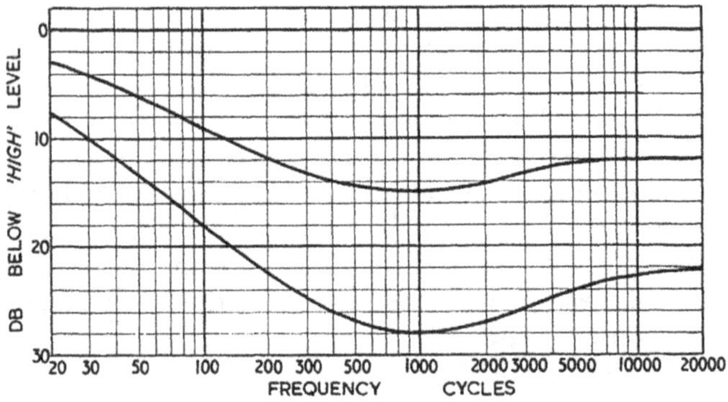

Fig. 910. *The contours produced by the circuit of Fig. 909.*

middle position inserts about 14- or 15-db attenuation and a corresponding amount of compensation. A lower position inserts approximately 25 to 30 db of attenuation to give low-level reproduction and provides a corresponding degree of compensation.

This approach is based on the philosophy that most people

usually want to listen at some specific order of level. Compensation is provided to suit these general orders of level. Finer adjustment can be achieved with the volume control and, if need be, slight adjustment of the tone control. Fig. 909 shows a schematic for this type and Fig. 910 gives typical curves. Choice here is a matter of individual preference rather than of engineering merit.

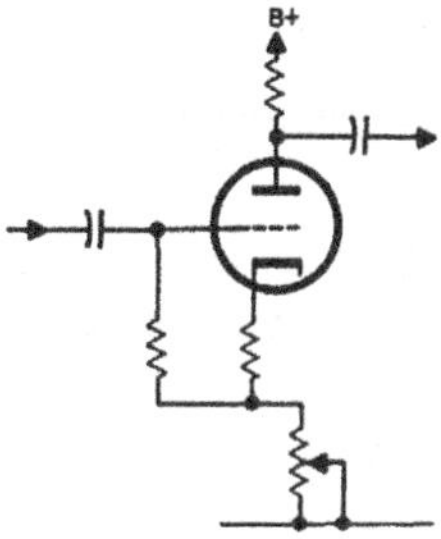

Fig. 911. *A single stage feedback circuit that gives a degree of volume control—but no "off" position.*

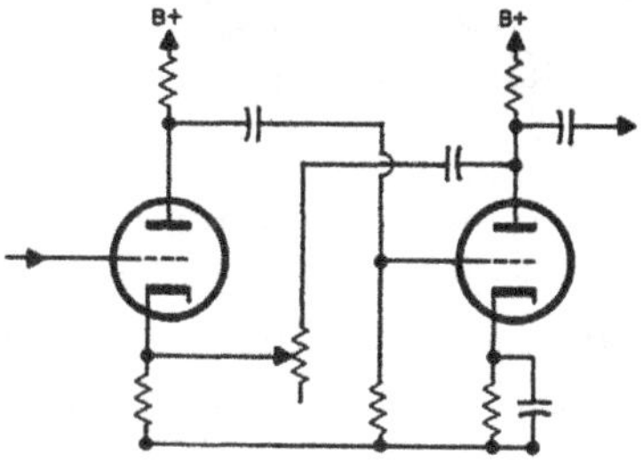

Fig. 912. *Using two stages gives a wider range of control, but involves design problems to maintain linearity.*

Feedback types

A straight (uncompensated) volume control can be achieved by the use of adjustable negative feedback (Fig. 911). Here adjustable feedback in the cathode circuit varies the gain of the tube. One disadvantage is that it does not provide an off position—only a limited range of gain control. It does have the advantage that negative feedback for the lower levels reduces distortion.

By using the feedback over two stages (Fig. 912) the range of the control can be increased somewhat. However, there is a problem here in producing a uniform flat response, due to the varia-

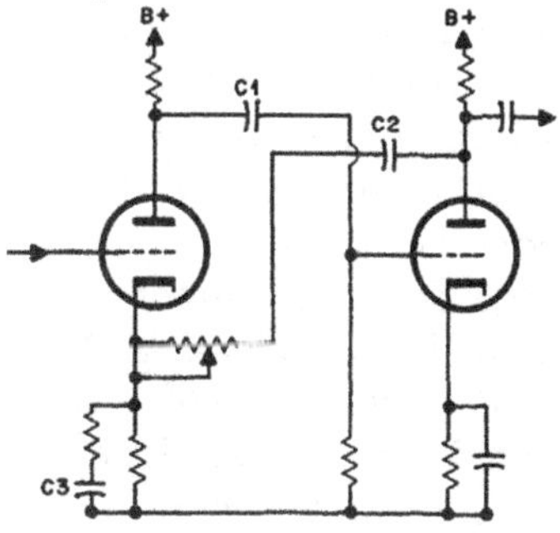

Fig. 913. *The two stage arrangement can be modified to give deliberate loudness contour variation.*

tion in feedback parameters. If these parameters are deliberately controlled (Fig. 913), a convenient form of loudness control is produced. Here the feedback capacitor C2 is responsible for the progressive bass boost as feedback is increased and gain reduced, while the small bypass capacitor C3 produces a progressive high-frequency boost up to 6 or 10 db, by choice of suitable components.

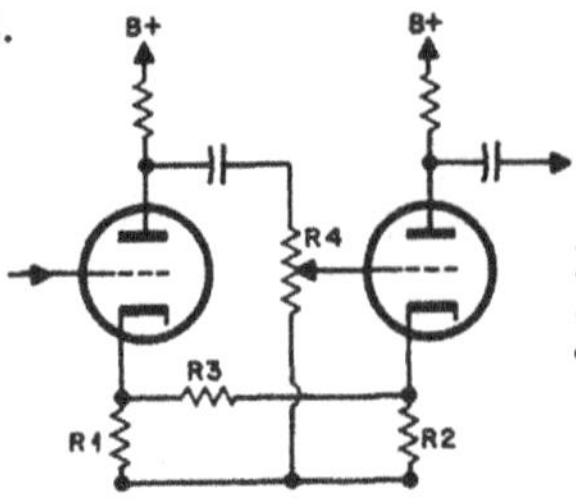

Fig. 914. *This circuit uses positive and negative feedback, as explained in the text, to extend the control range of the potentiometer; in this case as a straight volume control.*

Another feedback volume control (Fig. 914) uses positive and negative feedback. When the control is near its bottom end, R3, coupling the two cathodes, does not contribute any appreciable effect because the signal at the second cathode is considerably

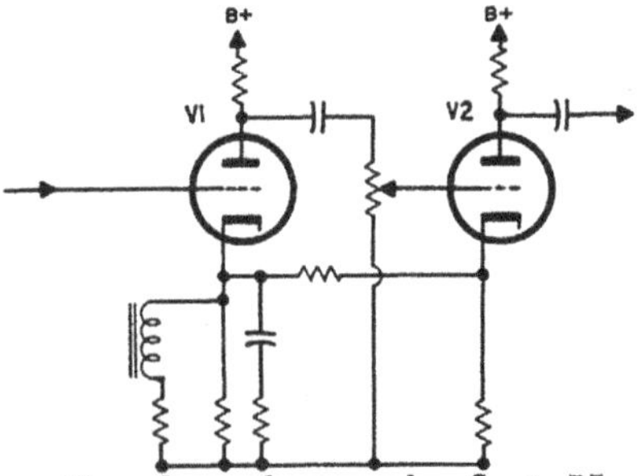

Fig. 915. *A modification of the circuit of Fig. 914 to incorporate loudness compensation functions. In this circuit the bias for V1 is principally determined by the resistance in series with the inductor (which may well be its dc resistance). The relationship between the remaining components determines the degree of compensation, and the frequencies at which it occurs.*

smaller than that at the first. However, R1 produces considerable degeneration in the first stage, maybe 6 to 12 db. As the slider is moved up the control to the top end, the signal at the second cathode increases and R3 produces positive feedback or regeneration, cancelling the negative feedback or degeneration in R1, due to the cathode current of the first stage. It may also provide regeneration beyond this cancellation, thus increasing the range of control provided by volume-control resistor R4 by as much as 20 db. The principal advantage is that it enables a wider range of gain control to be used with either a linear or semi-log potentiometer.

The only convenient way to make this circuit into a loudness control is by bypassing R1 with an additional inductor and resistance. By choice of a suitable component these can be incorporated into the one iron-cored coil. The circuit is shown in Fig. 915. The inductor and resistance provide a swing in the bass boost or rolloff in this case, while the capacitor and resistance do the same thing for the high end. With this arrangement the level response will probably be achieved somewhere below the top end of the control.

chapter 10

tone controls

THE function of a tone control used to be wider than in modern equipment. This was before the day of equalizers available with modern equipment to compensate for recording characteristics. At one time this had to be taken care of in the tone control.

Why tone control?

Any modern recording or radio channel has some tone-compensation system. This is to adjust program balance between high and low frequencies, to compensate for peculiarities in the studio reverberation characteristic or any other cause that may produce an apparent unbalance, excess or deficiency of high or low frequencies.

However, the home also has acoustic characteristics which lend coloration to the reproduction. Adjustment of tone balance is arbitrary because the factors that contribute toward the coloration are complex. The reverberation characteristic and frequency discriminating factors of both the recording studio and the listening room are usually far from being a smooth deviation that can readily be corrected by a simple tone-control system. As a result the best sense of balance in an overall system, taken from an original performance in the recording studio to the reproduction in a *particular* living room, must take into account the complicated acoustic characteristic of *both* rooms. The *overall* tone compensation needed should be a function of both the recording or broadcasting studio *and* the listening room.

The compensation put in at the recording or broadcasting end is adjusted to achieve the best sense of balance in the monitoring booth attached to the studio, usually made as acoustically "colorless" as possible without making it unnaturally "dead." It might seem, then, that all we need is a fixed tone compensation to suit the acoustic properties of the listening room in which the programs are heard and all programs should automatically come through with a good impression of balance.

This does not always produce the best results because acoustic colorations can interact. One studio may naturally emphasize the same groups of frequencies that our living room does, maybe at the high-frequency end. If the program from this studio is played without suitable adjustment to our tone compensation, this combination will make the overall effect seem excessively "live" over the range of frequencies where these peaks occur. Another studio, with the same average overall characteristic in this region, may tend to emphasize frequencies that would fall between those the living room emphasizes, with a resulting overall effect smoother than the average reproduction over the same system. Although, on some other criterion, such as a well-designed monitoring room, both studios might appear to need the same degree of tone compensation for the high-frequency end, in our living room the second program needs high-frequency accentuation to bring up the brilliance while the first one seems too brilliant and the high frequencies need playing down a little.

The modern trend in recording studios is to have a uniform reverberation time over the audio spectrum, considerably shorter than usual for an auditorium of corresponding size. Reverberation and characteristic coloration are then added by external means under careful control. This does produce program material basically within narrower tolerances, as regards apparent balance, and thus requiring practically no adjustment in any one living room between one piece of program recorded under these conditions and another under similar conditions.

While this may seem ideal, it is not always satisfactory. It gives what may be defined as high fidelity and does present a true impression of all the instruments. But some programs achieve some of their character due to the auditorium in which they are performed. Each of the famous ones, like the Albert Hall in London, has its own character due to the acoustic properties of the building. The acoustic background is as much part of the performance as the faint background contributed by the audience, an occasional cough, stirring of feet or applause. If the purpose of the recording or transmission is to give the listener the impression of being present at such a performance, removal of these incidental sounds and the general acoustic characteristic will destroy the illusion.

This leaves us with the fact that, even on modern high-fidelity wide-range recordings, it is desirable to have some degree of tone compensation for both high and low frequencies if we are to achieve the best impression of realism in our homes.

Other tone-control functions

In deciding what circuit to use, consider the tone control function. Sometimes it is called upon to get the best possible reproduction from an originally poor program source: for instance, an old 78 rpm recording that does not have any recorded frequencies much above 6 kc and is accompanied by considerable needle scratch or an AM radio transmission which has its frequency band limited by the number of channels squeezed into the broadcast band. The desirable bandwidth may vary from program to program according to the transmission conditions or the state of the record. It is desirable to have a flexible tone control that will make the best of the high frequencies present in the program without bringing out too much background noise with them.

A similar function is sometimes required at the low-frequency end, due to unwanted hum present on the program material. For example, some of the older recordings, made before the desirability of reproducing extremely low frequencies was recognized, may have considerable hum from the recording amplifier. In "antiquated" reproducing systems this hum was not noticeable. The system did not reproduce appreciably down to this frequency. However, the recording may have low frequencies in the region of say 80 to 100 cycles which we want to make the best of. So a flexible tone control that will achieve this may be helpful at the low-frequency end too.

If you want to pick off the audio from your TV set and play it through the hi-fi system, you will probably have to rework the audio section of the TV set, to reduce the hum level or else use a compromise setting of the low-frequency control to get satisfactory low-frequency reproduction without excessive hum. Most TV sets rely on the fact that their speaker doesn't reproduce much in the region of 60 cycles, hence have a smaller amount of smoothing (or other form of hum elimination) than necessary for a hi-fi system.

A control designed merely to vary the general level of the low-frequency end would need to be turned down so the whole program is noticeably bass-deficient before the hum is satisfactorily eliminated.

Other factors

These are the practical factors to consider in deciding upon a tone control circuit. The choice should also take into account the possible distortion that the amplifying stages associated with it can

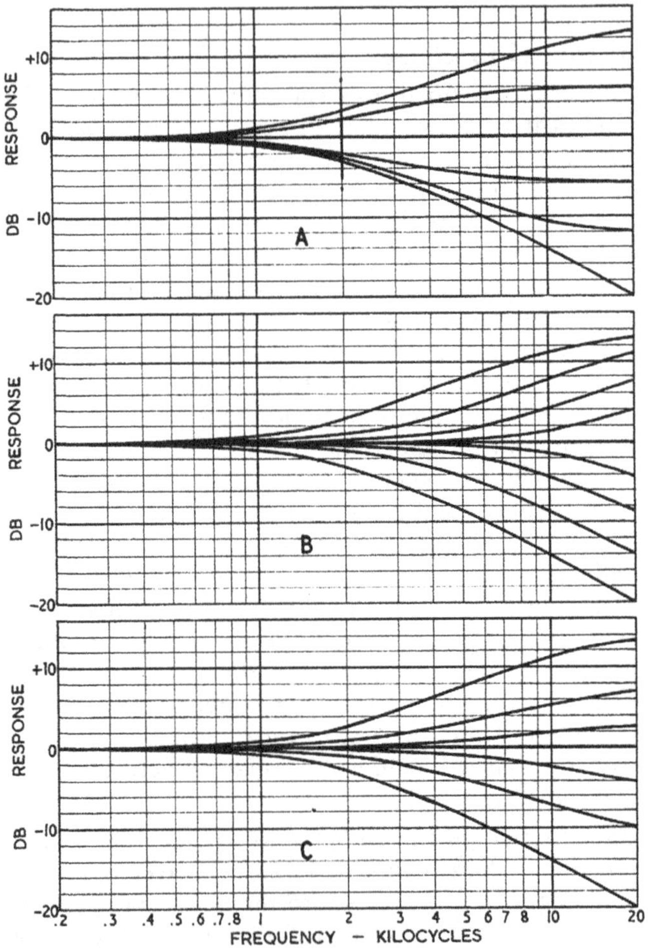

Fig. 1001. *Three ways in which high-frequency tone control, using simple R-C circuits, can be made to vary.*

introduce. It may inject different types of distortion since the circuit tends to distort one group of frequencies—the extreme low, middle or the high end, more than other regions. Also, consider the possible effect on IM as well as harmonic distortion.

The dynamic range is a factor in selecting a tone-control circuit, especially in deciding where to put it in a system. Any tone control needs to have some "headroom" if it is to provide boost at some frequency. It should be able to amplify, without producing distortion, considerably above the maximum level normal over the greater part of the frequency range. So the dynamic range of circuits incorporating tone controls needs to be somewhat wider

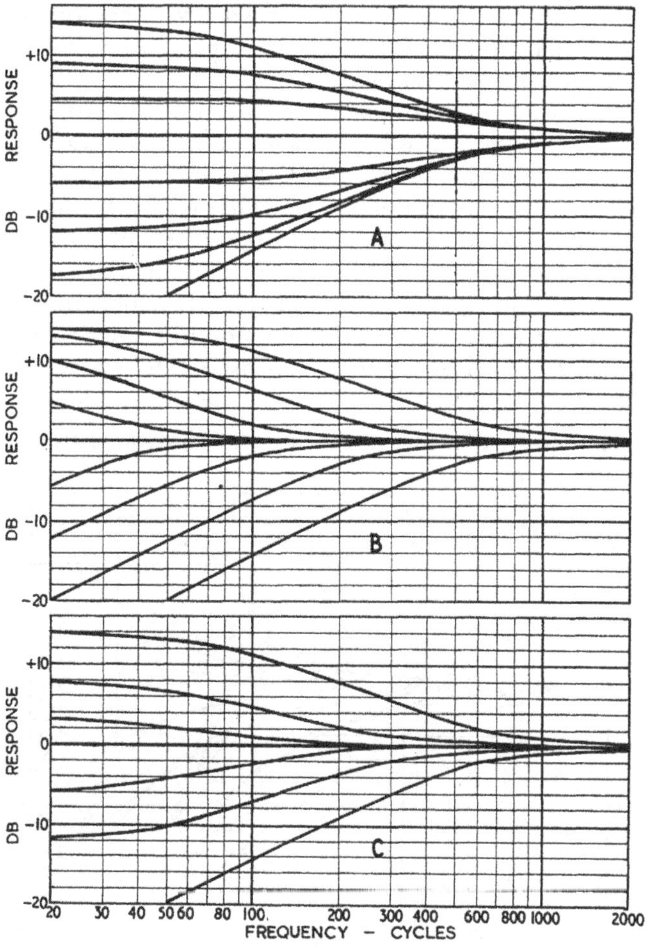

Fig. 1002. *Corresponding ways in which low-frequency tone control can vary.*

than that of stages used for straight amplification. For this reason it is good to place the tone control after the major volume control so the level of the program material passing through the tone control is reasonably constant compared to the reproduced level. This will avoid requiring excessive dynamic range to take care of the margin of boost required and the variation in program level from the different sources (and the different listening levels required as well).

Response form

The simpler tone controls utilize resistance and capacitance elements that produce a kind of step response similar to that from which equalizers are built. Step responses are used for effective boost while capacitances in a position to produce rolloff reduce the level of low or high frequency.

But circuit response can be changed in more than one way by the adjustment of the control. Fig. 1001 shows basic ways of varying the high-frequency response. At A the *amount* of boost or rolloff is adjusted, but the frequency at which either begins to take effect is constant. At B the amount of boost or rolloff is constant but the *frequency* is shifted. At C the response variation goes between the two; both the amount and the frequency are changed to produce a smooth transition from maximum boost to maximum rolloff with almost proportionate variation all the way, above about 2,000 cycles.

Fig. 1002 shows the same variations and characteristics applied to the low-frequency end.

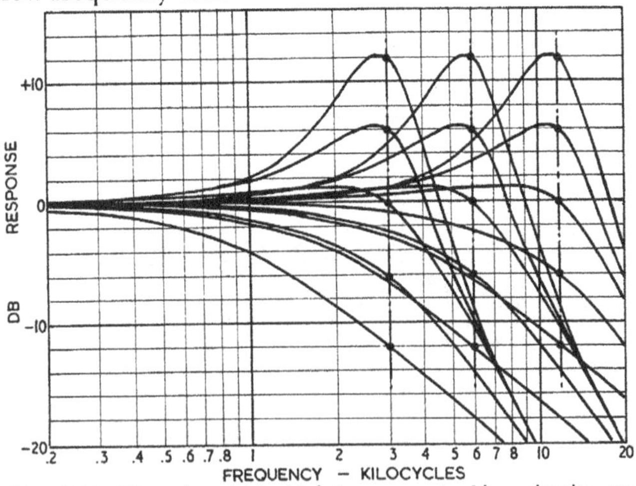

Fig. 1003. *Ways in which high-frequency peaking circuits can change their response. Curves using the same critical frequency are linked by the dot-and-dash lines. Usually resistance varies height or rolloff, while changing L or C alters the critical frequency.*

For simpler controls aimed at correcting for slight deviations in good high-quality modern recording or transmission, circuits providing this kind of response variation are quite adequate. For controls intended to make the best of poorer conditions, peaking

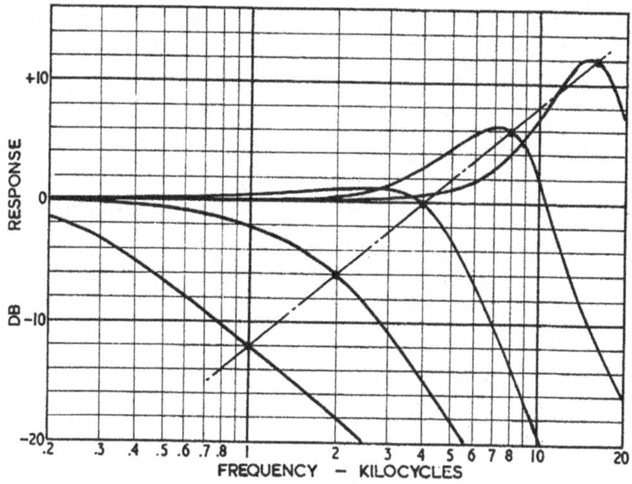

Fig. 1004. *When peaking is varied by using feedback over a two-stage amplifier, this kind of frequency response change can be expected.*

and vari-slope type circuits are sometimes advantageous either as an alternative or an additional means of control.

There are two ways of producing both varieties of circuit. One

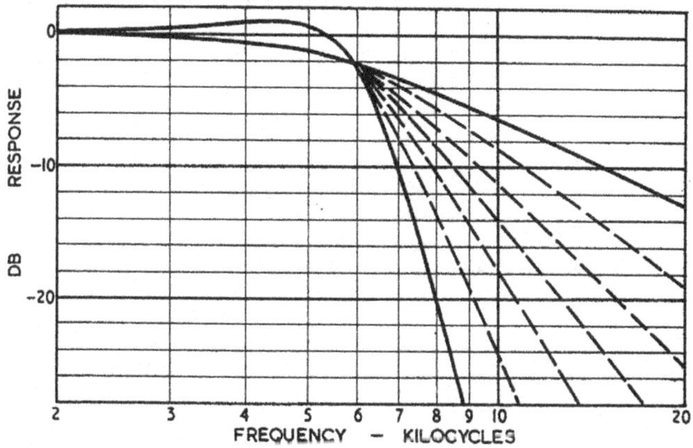

Fig. 1005. *Variable slope rolloff is sometimes a useful feature.*

uses inductances as well as capacitors while the other employs feedback to achieve a similar kind of interaction. The peaking response can be varied in several ways as in the step circuit. The critical frequency and also the amount of peaking can be adjusted (Fig. 1003).

Using inductances and capacitances together, the *frequency* of the peak or rolloff is determined by the combined values of L and C, usually fixed or adjustable in steps, while the *amount* of peaking or rolloff is adjusted by the resistances. Another variation changes the amount of peak and the frequency at the same time (Fig. 1004). This is more conveniently produced with the feedback type of control, using only resistors and capacitors.

The remaining type uses a variable-slope rolloff. For this there are two approaches also but both use a fixed frequency, determined by the reactance components, and adjust the slope of the rolloff by damping or partially bypassing the overall reactive phase-shift effect. The frequency response variation for one particular frequency is shown in Fig. 1005.

These are the responses that can be produced. A complete arrangement can consist of any one or two of these controls or a combination of as many as desired.

Simple step circuits

Fig. 1006 shows the simplest possible arrangement of a high-

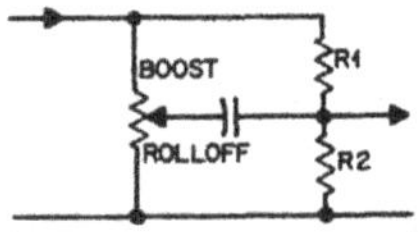

Fig. 1006. *The simplest circuit for making high frequency tone control, including boost and rolloff.*

frequency control giving rolloff and boost: two resistors, a capaci-

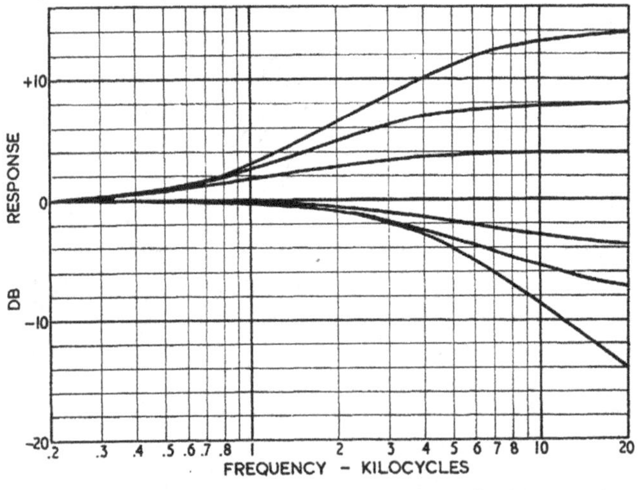

Fig. 1007. *When the circuit of Fig. 1006 uses R1 having four times the value of R2, this is the response variation produced.*

tor and a potentiometer. If the resistors are of equal value, the maximum boost is 6 db. Usually a bigger boost than this will be desirable as a maximum. But, with this particular value, the boost and the rolloff frequencies at maximum and minimum settings coincide. If larger resistors are used for R1 and R2, say a ratio of 4 to 1, to give 14-db maximum boost, the frequency at which the maximum boost starts to take effect will be one-fourth the frequency at which maximum rolloff begins (Fig. 1007).

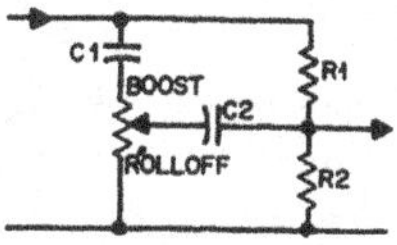

Fig. 1008. *The addition of an extra capacitor (C1) overcomes the unbalance shown in Fig. 1007, to some extent.*

This can be equalized to some extent by putting a capacitor in series with the top end of the potentiometer so that, at maximum boost, there is an additional capacitor in series with the one connected between the slider and the junction of the fixed resistor. This means the effective capacitance in the circuit changes according to the position of the slider on the potentiometer.

One desirable feature about any tone control is that it should be possible to obtain a flat frequency response, preferably at some setting of the variable control, so adjustment can be taken right through flat. Otherwise the only way to achieve flat for comparison is to switch the tone control out of the circuit altogether.

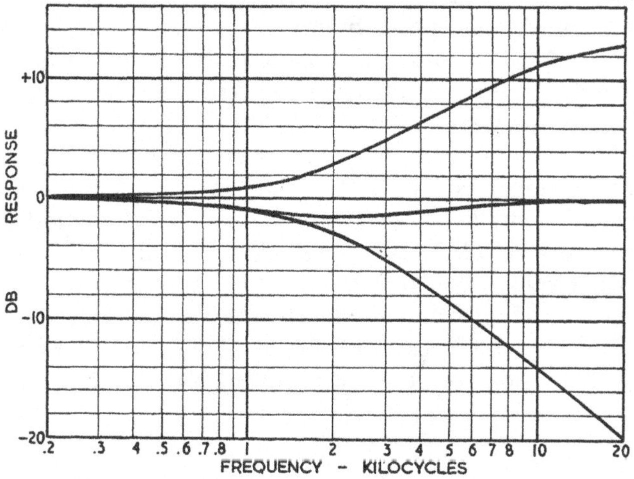

Fig. 1009. *The circuit of Fig. 1008 can never give a level response at any position. The nearest approach is seen here.*

With the circuit of Fig. 1008 the nearest approach to flat response occurs when the potentiometer slider divides its resistance in the same ratio as the fixed resistors. Then the voltage division at mid-range frequencies and at the ultimate high-frequency end is identical, since the capacitance in series with the top end of the potentiometer changes progressively from open to short circuit. But at a frequency where the capacitive reactance is approximately equal to the resistance of the potentiometer there is a 45° phase difference between the open-circuit voltages at the junction of R1, R2 and the potentiometer slider. The latter voltage is lower than the other by 3 db. Connecting the capacitor between the two points will make a dip in the overall response, dependent on the relative value of the resistances for the fixed and variable components. (Fig. 1009).

An advantage of this circuit, compared to the next one, is that it is possible (by controlling the value of the capacitor in series with the top end of the potentiometer) to use a boost frequency different than the rolloff frequency, which is controlled by the value of the capacitor in series with its slider, and still achieve an approximately similar approach to level response in the middle position.

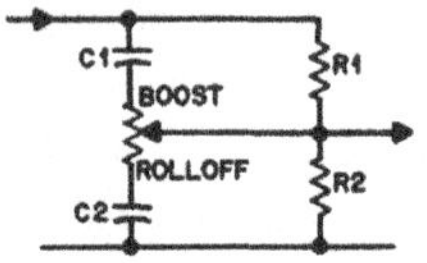

Fig. 1010. *This circuit, using symmetrical choice of components, can give the range of responses shown in Fig. 1001-A.*

Provided all the values are in the same proportion, the circuit of Fig. 1010 produces a flat response at a correct middle setting. Also, the frequency of maximum boost is the same as that where rolloff takes effect. Thus it achieves the response variation of Fig. 1001-A. To do this the ratio of R1 to R2 must be the same as that of C2 to C1. In other words, the *reactance* of C1 to C2 should have the same ratio as R1 to R2. The point of rolloff or commencement of boost is where the reactance of C1 is equal to R1 and the reactance of C2 to R2. Because of its ability to produce a uniformly flat response at one position of the control, this circuit is quite popular. It is often used with the values divergent from this ratio to vary the boost and rolloff combination obtainable. An intermediate position, then, can never produce a flat response.

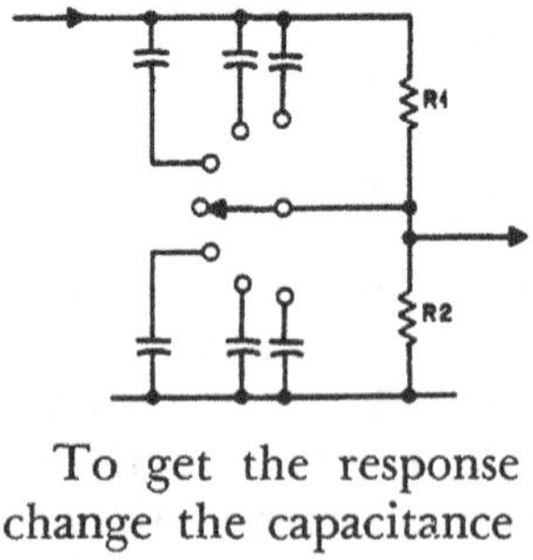

Fig. 1011. *A switched circuit that gives the response variation of Fig. 1001-B.*

To get the response of Fig. 1001-B, the only method is to change the capacitance values, because the constant boost value means the circuit resistance values must be constant. The only practical technique is to use a switched arrangement such as that in Fig. 1011.

A fairly comprehensive tone control can be made by combining the two. A simple way of doing this uses a variable resistance to

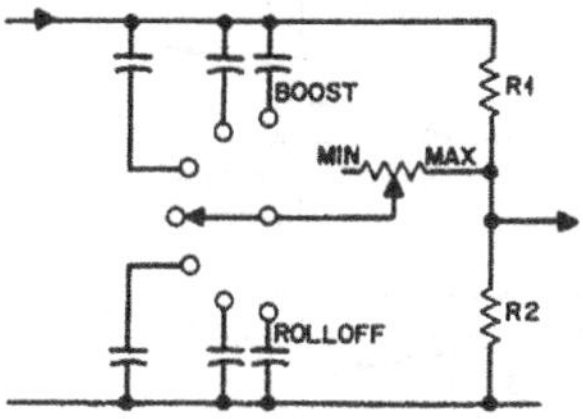

Fig. 1012. *Addition of a variable resistance gives greater versatility to the circuit of Fig. 1011.*

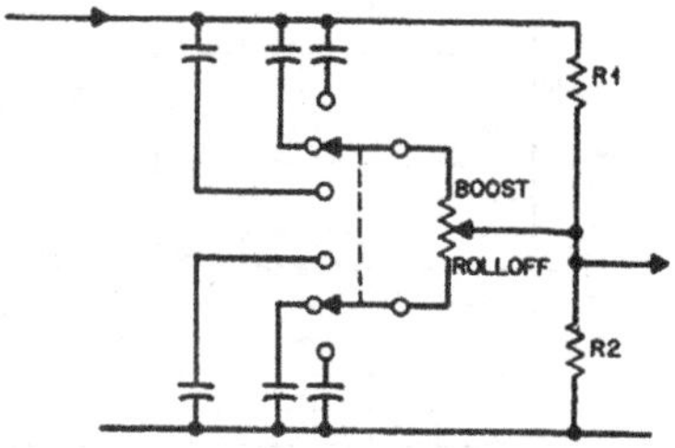

Fig. 1013. *Using a two-pole switch, a response variation combining the features of 1001-A and -B can be produced.*

adjust the amount of boost or rolloff by putting the resistor in series with the switch slider (Fig. 1012). But this does not give adjustment *through* flat. There is a flat position but it is isolated from the boost and rolloff positions. An alternative combination that uses a switch to select frequency and a potentiometer to adjust amount of boost *or* rolloff, and which does go through flat, is shown in Fig. 1013. It achieves a variation of the *frequency* of boost or rolloff by the switch (as in Fig. 1001-B) while the *amount* of boost or rolloff (Fig. 1001-A) can be adjusted by the potentiometer. In this way a considerable variety of responses can be obtained.

There are advantages in using two controls to get greater versatility, but corresponding disadvantages. The fact that two knobs have to be adjusted to achieve a desired response (only high-frequency response, remember) means that it is less likely that the best response will be obtained. A single knob that gives a range of characteristics of better average desirability, makes the control simpler and reduces the chance of seriously incorrect settings.

Fig. 1014. *A high frequency control using a two gang potentiomenter that produces response variation of the type, Fig. 1001-C.*

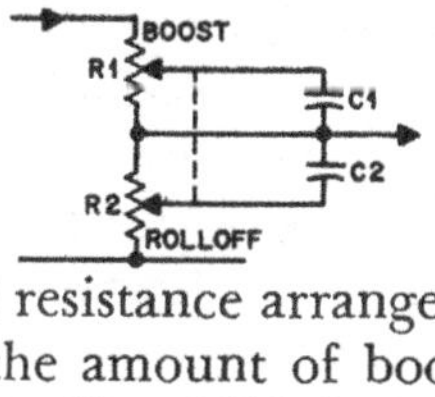

Fig. 1014 shows a circuit using a ganged resistance arrangement that varies both the turnover point and the amount of boost or rolloff, producing a response variation as in Fig. 1001-C. R1 and R2 are potentiometers. While a reasonably good control can be made using linear potentiometers, there is an advantage in semi-log or full log tapered controls. One should be inverted taper so that, at a middle setting of the control, each capacitor is shunted by about one-tenth of its resistor, while about nine-tenths of each resistor is in the circuit. Fig. 1015 shows the ranges of control, us-

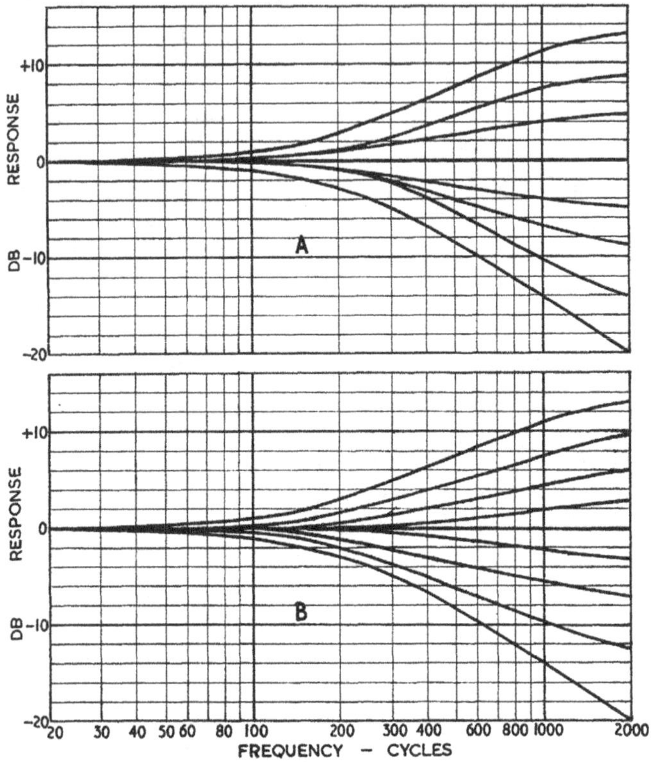

Fig. 1015. *Showing the effect of controls using different tapers in Fig. 1014: A is the variation using linear taper, B with log taper, complementary (one direct log, one inverse log).*

ing a pair of linear potentiometers (A) and a pair of complementary logarithmic potentiometers of standard taper (B).

Turning to arrangements to produce a straightforward bass boost or rolloff, Fig. 1016 shows the only way to vary bass rolloff directly. This necessitates changing the coupling capacitor for each rolloff required. Using a resistor in shunt with the rolloff capacitor produces the same rolloff point but limits the rolloff to a degree of stepdown. If a coupling capacitor is required to produce dc separation, as when coupling from a plate to a following grid, two capacitors should be used: a larger one to provide dc blocking while the smaller one for variable bass cutoff has the variable resistor in shunt with it.

If the variable resistance goes completely open-circuit, the rolloff is that provided by the capacitor but, as soon as some resistance is in shunt with the capacitor, the rolloff is limited to a stepdown response. Short-circuiting the capacitor produces a rolloff at the position determined by the main coupling capacitor, which for the purposes of tone control design should be below the audio-frequency range.

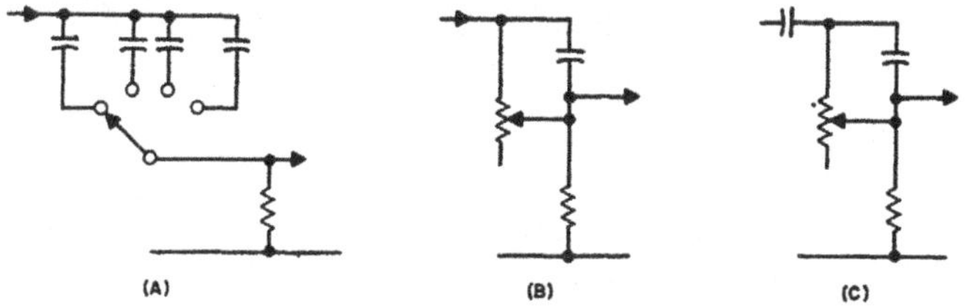

Fig. 1016. *How bass rolloff is effected: circuit A selects the rolloff frequency by switching; circuit B modifies the rolloff to a step-down of variable depth; circuit C includes a dc blocking capacitor.*

Bass boost is provided by a capacitor in the bottom end of an attenuation arrangement (Fig. 1017). This is similar to the arrangement for equalizer circuits. The amount of boost can be adjusted by using a variable resistor directly across the capacitor. When the capacitor is short-circuited, there is no boost. Maximum boost is secured by making the resistor a high value, approximating an open circuit, for the lowest frequency in the audio band.

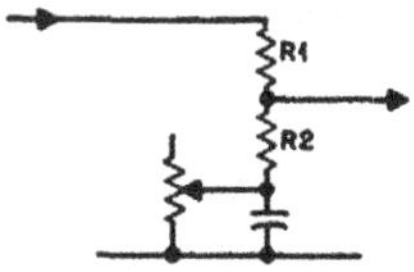

Fig. 1017. *A simple adjustable bass boost circuit.*

Fig. 1018 combines the rolloff and boost arrangements to use one variable resistor for the response of Fig. 1002-A. The value of the potentiometer should be very high compared to either R1 or R2, unless an appreciable portion of the boost or rolloff is to be sacrificed by its presence.

This is a detrimental aspect of the circuit because it often dictates the need for a potentiometer of several megohms to avoid deteriorating the maximum boost condition. Usually the grid of the following stage has a maximum permissible circuit resistance, which makes this value impossible. So a compromise has to be accepted.

However, a mitigating factor is that the loudness contours of human hearing converge at the low-frequency end. Consequently a certain amount of bass boost is more effective than the same amount of treble boost. This means that some of the available boost can successfully be sacrificed by using a lower-than-ideal value for the potentiometer of Fig. 1018.

The frequency where the boost or rolloff begins to take effect

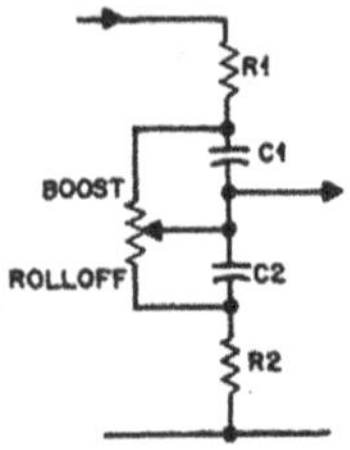

Fig. 1018. *The commonest bass, or low-frequency tone control circuit. It gives the response variation of Fig. 1002-A.*

is determined by making the reactance of C1 equal to R1 and the reactance of C2 equal to R2.

Fig. 1019 shows an arrangement that gives switched control to achieve the response variation of Fig. 1002-B. This requires separate capacitors to give each position of boost or rolloff.

For the circuit of Fig. 1014, producing the combined response of Fig. 1001-C, the important part of the potentiometers, affecting both the frequency and the amount of boost or rolloff, is that *in parallel* with the capacitors, or the part from the slider to the center point of the circuit. The other section of each potentiometer serves only to pad out the circuit so the mid-band attenuation is kept constant as the control is turned. The only way to get the same kind of variation in response shaping at the low end of

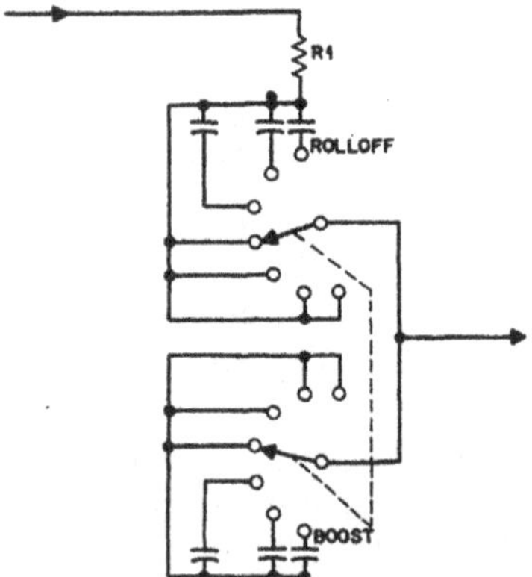

Fig. 1019. *A switched tone control circuit giving the response variation of Fig. 1002-B.*

the range is to use complementary variable resistances *in series* with a pair of capacitors. The relative values of the pair of capacitors will then determine the attenuation at the extreme low-frequency end of the range while the attenuation in the center part of the range will be secured by the relation between the two components of series resistance. (Fig. 1020).

While this will produce the kind of response variation of Fig. 1002-C, it will also vary the apparent gain of the amplifier drastically. At maximum boost the gain will be almost zero while at maximum rolloff the gain will be almost the full gain of the stage, without any attenuation in the mid-band region.

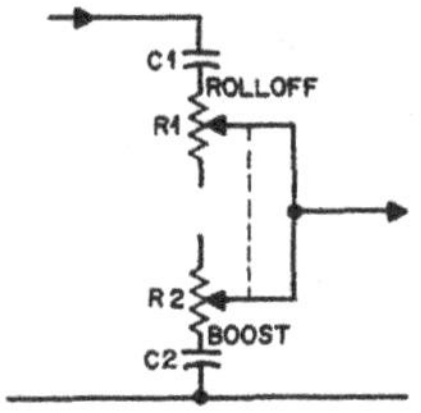

Fig. 1020. *A low-frequency circuit producing the variation of Fig. 1002-C. This changes mid-band gain drastically.*

The circuit of Fig. 1021 overcomes this to some extent. The maximum boost is limited by putting a fixed resistor R2A in series with R2 so the mid-band attenuation at maximum boost is limited. A third potentiometer is ganged to these two to offset the change in mid-band attenuation.

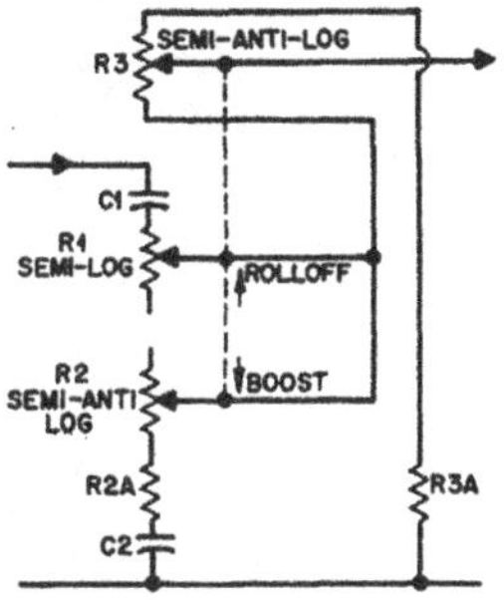

Fig. 1021. *Addition of a third variable to the gang of Fig. 1020 helps overcome the gain change. But it still is not constant.*

To get the optimum effect, the best taper to use is complementary semi-log. This will give the best approach to the constant mid-band gain and a maximum variation in the characteristic. Use of a full-log taper is likely to produce much more variation with gain as the setting is adjusted. It will also require a considerably greater overall attenuation, but it will give a greater range of control. But, whatever type of control is used, the mid-band gain does not stay truly constant as it does with the corresponding high-frequency control. This method of adjustment is scarcely recommended for the low-frequency end for this reason as well as the fact that it is more expensive because it requires a three-gang rather special type of potentiometer.

Practical considerations

A typical complete circuit needs to combine the step-boost and rolloff responses. Fig. 1022 shows two straightforward circuits that do this. In applying these to a practical amplifier certain factors need attention. They need to be fed from the plate of a preceding stage and also to connect to the grid of a following stage. Also the grid-circuit resistance of the following stage must not exceed the permissible value at maximum bass boost.

In calculating the overall response available remember that the plate circuit impedance of the preceding stage modifies the calculations. The plate circuit resistance is, of course, the combined resistance made up of the coupling resistor in parallel with the

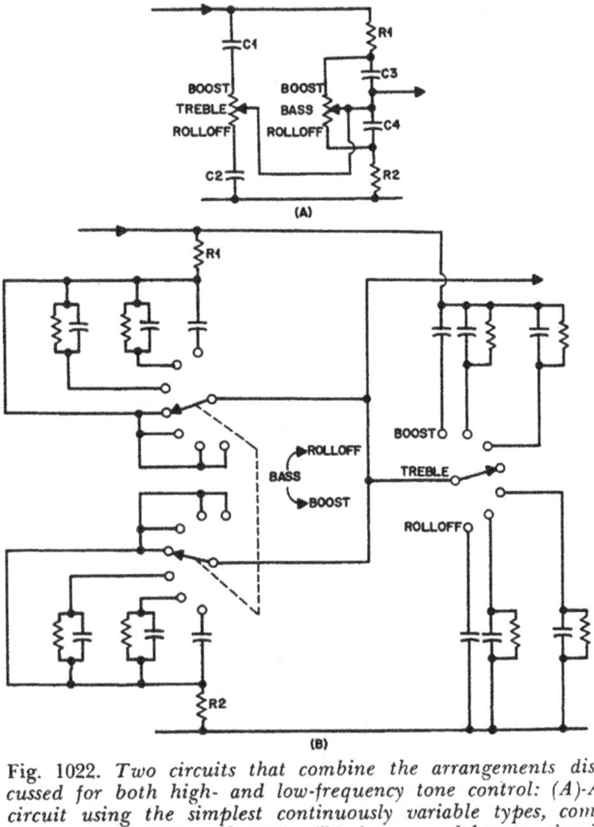

Fig. 1022. *Two circuits that combine the arrangements discussed for both high- and low-frequency tone control: (A)-A circuit using the simplest continuously variable types, combining Figs. 1010 and 1018; (B)-A more elaborate circuit using switching to produce variations as at Figs. 1001-C and 1002-C.*

ac resistance of the tube. This will limit the maximum high-frequency boost because the capacitor shunting the upper end of the attenuator does not bypass this source resistance so there is still some attenuation at the extreme high end where, according to the idealized form, the attenuation should be short-circuited by C1.

Usually the source resistance will not have too much effect on the available bass boost and rolloff. What limits the available bass boost and rolloff is the value of the control resistor used for this purpose. This is determined by the permissible grid-circuit resistance for the following stage.

By careful choice of values, it is possible to achieve an equal degree of boost at both ends of the frequency response. An alternative choice of values can produce any combination of available boost for the system in hand.

With all these circuits the available amplification with maximum boost will be that much more than in the mid-frequency range. This, of course, should be an obvious statement. It means, however, that there must be this much attenuation in the middle of the band which is returned to some extent by the boost circuit.

Some disparage the use of the word "boost" because they contend it implies the gaining of something for nothing. However this is not the intention here. The word boost is considered more satisfactory than any other term because we really consider the response *relative to the apparent loudness in the middle of the band*. If the apparent gain of an amplifier is greater at the extreme low- or high- frequency end than in the middle frequencies (which determine the apparent loudness of the program material) then the audible effect is that the low frequencies or high (as the case may be) are *boosted,* compared to the general loudness level.

The general loudness level may have to be attenuated, and the extreme low or high frequencies not, and thus the term boost may be regarded as *academically* incorrect. However, it is more acceptable than the thought of attenuation, rolloff or cut, which does not convey an accurate idea of its effect on performance. Definitely, bass boost cannot be regarded as synonymous with treble cut, and vice versa, as some sticklers for "accurate terminology" insist!

The tone-control circuit should be so placed that distortion does not occur at any position. If boost is required, the implication is that these frequencies are deficient and thus increased amplification of these frequencies is unlikely, in itself, to cause distortion. So it is reasonably safe to design a circuit on the basis of the levels in the mid-band. This may not always be true at the low-frequency end, however, where boost may be employed to make the frequencies audible, even though the power level may be fairly high.

Increased sensitivity of the human ear at the high-frequency end means that the energy, and hence voltage level, even when boost is applied, will still be relatively low. Application of boost to a program with appreciable energy at the high-frequency end would be most distressing to listen to.

An alleged disadvantage of the tone control circuits we have discussed is that distortion is cumulative; that is, it is not possible to apply feedback to reduce distortion. The only way to minimize distortion is to design each stage for minimum distortion as a straightforward amplifier. Of course it is possible to apply a degree of feedback over that stage by using cathode degeneration.

If the maximum boost is of the order of 14 or 15 db, (usually considered acceptable) then the stage before or after need have no more gain than this, to replace the loss in the network. This means that some gain can usually be sacrificed to linearize the amplification by cathode degeneration. A large cathode bias resistor, one that is a considerable fraction of the plate coupling resistor value, will produce this degeneration. The bias resistor is not bypassed. This reduces the amount of distortion produced in this stage although there is a limit to its effect: use of too large a bias resistor will overbias the tube. So there is an optimum choice for the circuit.

To improve the design still further a cathode follower can be utilized to eliminate the source-resistance problem. Alternatively, two stages of amplification can give a very considerable gain—between the preceding volume control and the tone control circuit. This provides an opportunity to use a large amount of negative feedback to reduce distortion due to the curvature of the tubes.

Feedback types

The disadvantage of this combination is that gain has to be thrown away twice: once to minimize distortion in the amplification and again to provide headroom for the low- and high-frequency boost. It was this thinking that gave rise to the introduction of feedback-circuit tone controls (Fig. 1023). Instead of controlling the frequency response in the forward gain of the amplifier, between stages, the frequency-selective network is placed in the feedback.

In basic theory the amplification has no frequency discrimination; only the feedback does. In feedback theory, if the loop-gain factor $A\beta$ is large compared to 1, the overall gain of the amplifier is approximately equal to $1/\beta$. In other words, it is almost independent of the internal gain of the amplifier, A, and principally dependent on the feedback fraction, β. So a simplified explanation of the action of feedback tone control says that we place a frequency-discriminating network in the feedback section and thus produce an inverse response for the overall amplifier gain.

Unfortunately, this approximation is based on simple algebra, which takes no account of phase *shift*. Using algebra, without incorporating the operator j, the feedback seems to be either positive

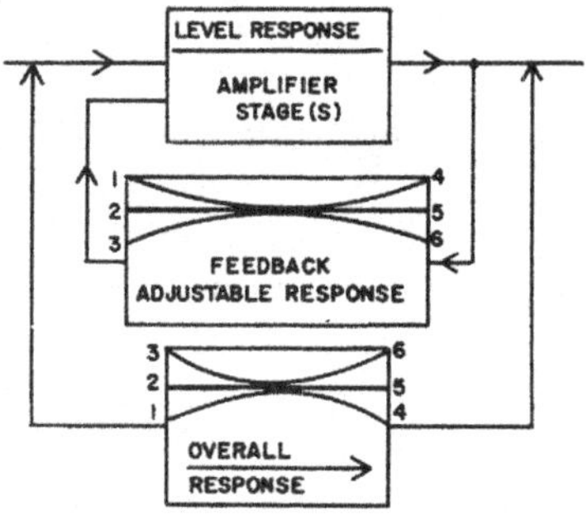

Fig. 1023. *Illustrating the basic concept behind feedback tone control.*

or negative. We can take the feedback factor as *either* $1 + A\beta$ *or* $1 - A\beta$. In practice $A\beta$ is a complex quantity involving a phase angle, particularly when frequency-discriminating elements are included to produce a tone-compensation arrangement. Then $A\beta$ is neither positive nor negative but a complex quantity which may be regarded as something in between, but still a large quantity, a long way above zero.

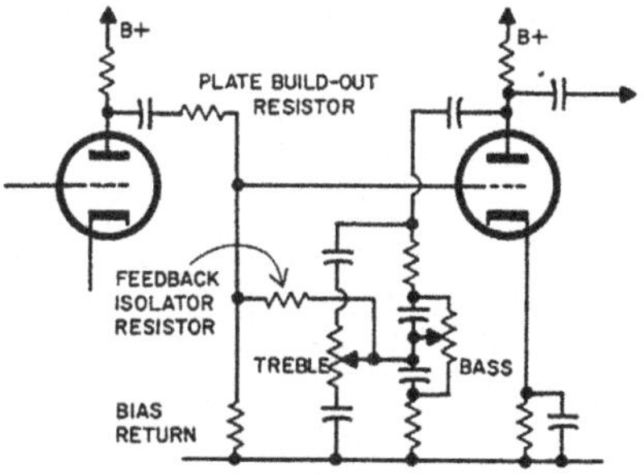

Fig. 1024. *A possible circuit utilizing feedback over one stage.*

In the case of equalizer circuits, the design can be taken in stages and a satisfactory overall response achieved. With a tone-control circuit, where *variation* in response is required, a little more care is necessary because the interaction must produce the right kind of variation in response at all points.

There are basic limitations to feedback tone controls. The maximum *range* is set by the gain of the section over which feedback control is applied. Suppose the section gain is 26 db. Feedback can cut the mid-band gain by, say 14 db, to 12 db. Progressive removal of feedback can then give 14-db boost. But progressive increase of feedback to $\beta = 1$, will give only a maximum cut of 12 db. So the maximum difference between boost and cut is the section gain, in this case 26 db. Some overcome this by using feedback only for boost, then using a simple cut circuit elsewhere for the other part.

Feedback over just one stage, from the plate to the grid, can be used (Fig. 1024). Here a resistor is interposed between the plate of the preceding stage and the feedback point to permit adequate resistor values to be used for the feedback network and to prevent the preceding stage from causing distortion due to the low effective resistance load presented at some frequencies by the feedback arrangement. This uses essentially the same circuit configuration as the forward control network of Fig. 1022-A, except that, instead of feeding the following stage, it feeds back to the grid of the same stage.

Thus the feedback network inverts the response achieved. As there are no additional coupling reactance elements other than the single coupling capacitor, it is quite possible to achieve a satisfactory control response range within the limitation of stage gain. It is necessary, however, to insure that the components that adjust the feedback response do not appreciably modify the forward gain characteristic due to the loading effect on the previous stage coupling; otherwise the tone-compensating effect is somewhat nullified. That is the reason for the additional resistors in the circuit.

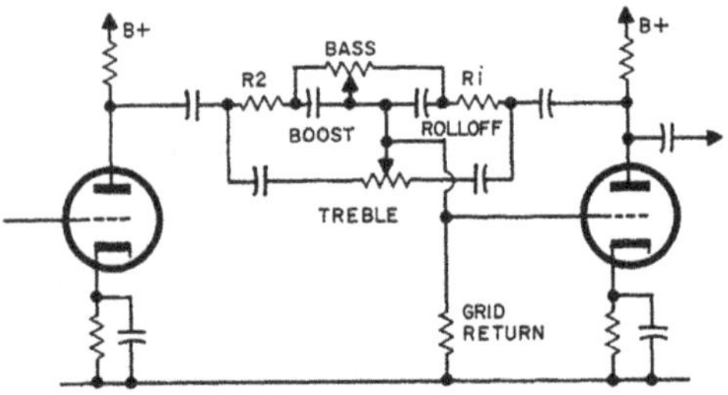

Fig. 1025. *An improvement on the circuit of Fig. 1024 that makes the circuit double-acting in its tone control effect.*

By using the plate-circuit feed from the previous stage as the "bottom" end of the feedback element, as far as frequency discrimination is concerned, the control becomes double-acting. Half its effect is due to a change in coupling between stages and half to feedback (Fig. 1025).

By using two stages (Fig. 1026), the feedback can be taken from the plate of the second stage to the cathode of the first stage. This serves to isolate the effect of the tone-compensating components from the feed from the preceding stage, which was a problem in the circuit of Fig. 1024. However, the fact that there is an additional stage, with its coupling capacitor and stray capacitance, means that the response contributed by these, (particularly phase

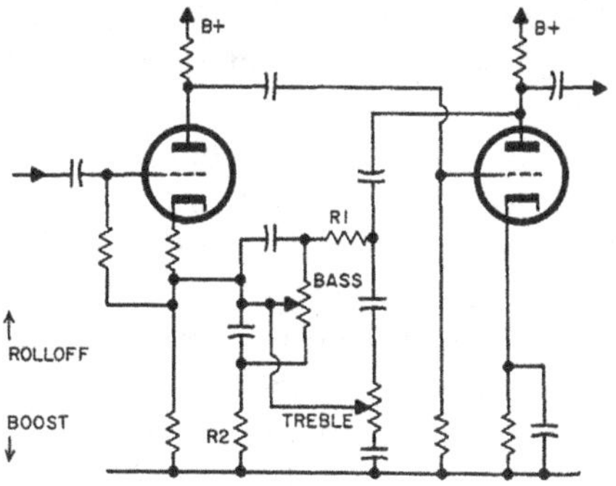

Fig. 1026. *A feedback tone control, using two stages, from plate of the second to the cathode of the first.*

effects in the forward part of the amplifier) interacts with the response of the feedback arrangement. This must be taken into account in designing the overall performance.

A well-designed feedback tone control does have the advantage, particularly with regard to low-frequency boost, on the score of distortion. The maximum amount of feedback is attained for the greater part of the frequency range. If more low-frequency response is required, some of the feedback is removed to achieve the necessary low-frequency boost. This means that the distortion of the low frequencies may be somewhat increased in comparison to mid-range frequencies. Won't this introduce intermodulation, too?

Fortunately, the fact that the mid-range frequencies employ a greater amount of feedback means that intermodulation effects, normally introduced in an amplifier at low frequencies without feedback, will be minimized by the feedback *according to the amount available at the middle frequencies*—the ones which would be modulated. The fact that the curvature may introduce fairly low-order harmonic distortion of the low frequencies is itself practically impossible to detect. We usually hear the extremely low frequencies principally by their harmonics anyway. So a little more or less of second-harmonic distortion of the lowest frequencies is comparatively unimportant, provided it is not accompanied by the much more irritating intermodulation distortion that usually accompanies harmonic distortion.

This is where the feedback tone-control arrangement helps particularly by minimizing the intermodulation distortion, although it may not reduce the harmonic distortion to the same extent.

For the dynamic range situation, the feedback tone control is quite similar to the direct type and must be treated accordingly in selecting the point in the amplifier at which to apply it.

Peaking types

The simplest circuit for peaking control can, in theory at least, be built around an interstage or some other type of transformer. Low-frequency peaking can be produced by resonating the coupling capacitor with the primary inductance of the transformer.

The degree of peaking or rolloff can be adjusted either by varying the source resistance, which in this case consists of the plate resistance of the tube and the coupling resistance in parallel, or by varying a resistor in parallel with the primary winding. The former is not a very suitable form of adjustment so the latter is usually chosen in this case. The higher the resistance the greater the peaking. At the same time, using a lower value of resistance (which will not necessarily attenuate the mid-band seriously) will damp the peaking and produce a rolloff. Suitable choice of values permits a good range of control to be achieved without appreciable change in mid-band gain (Fig. 1027).

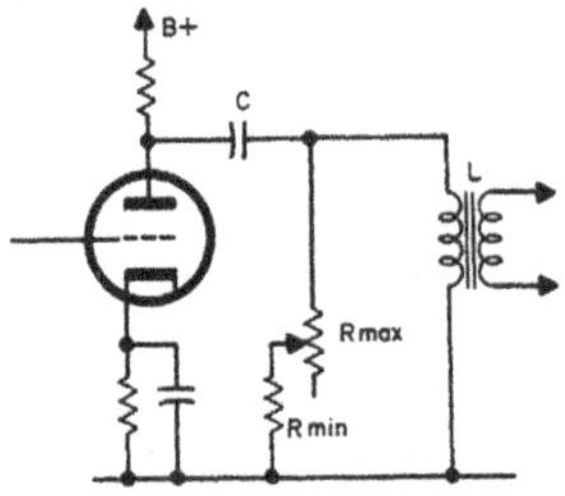

Fig. 1027. *The part of an L-C circuit that produces control of low-frequency peaking. Inductance and capacitance control the critical frequency; the value of R across the primary controls the peaking or rolloff.*

For high-frequency peaking the leakage inductance is resonated with the transformer secondary capacitance—either the natural capacitance or some artificially added. This can also be adjusted for the height of peak by variable resistance in series with the primary or in shunt across the secondary. If an additional separate capacitance is used to provide the resonance effect, an alternative position is in series with the separate capacitance across the secondary. These possibilities are illustrated in Fig. 1028 and the possible responses achieved in Fig. 1029.

To get a reasonable degree of independence between the two controls so the low-frequency control does not affect the high-frequency control and vice versa, a special design of transformer is required so that the parameters are well separated and the effective values of resistance, on the primary to control the low-frequency peaking and on the secondary for controlling the high-frequency peaking, are quite different, taking into account any stepup or stepdown ratio of the transformer.

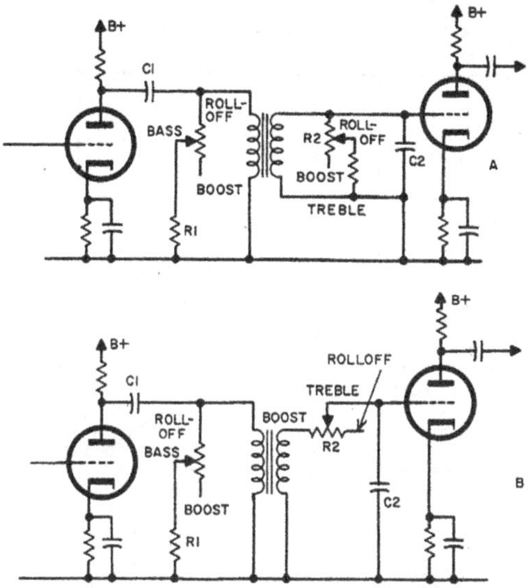

Fig. 1028. *Two ways of adding high frequency control to the low-frequency control of Fig. 1027*

As transformers are little used for interstage purposes these days and the acquisition of a special type would prove difficult, this item is presented here only for the sake of completeness. It is a possible method of control that has actually been used. It illustrates the difference between this kind of control and a somewhat similar one produced by the feedback arrangement.

If the amount of feedback over a two-stage R-C-coupled amplifier is varied, the degree of peaking will alter according to the amount of feedback applied (Fig. 1030). Not only the amount of peaking but also the frequency of the peak will shift. In the case of low-frequency peaking, the effect of the feedback is to extend the frequency range farther as a greater amount of feedback is applied. This means that as the peak becomes higher its frequency becomes lower. At the top end of the range, if this method is applied, making the peak higher will also raise its frequency.

Fig. 1031 shows a circuit employing a 6AW8 using this principle for the low-frequency end. The amplification changed for feed-

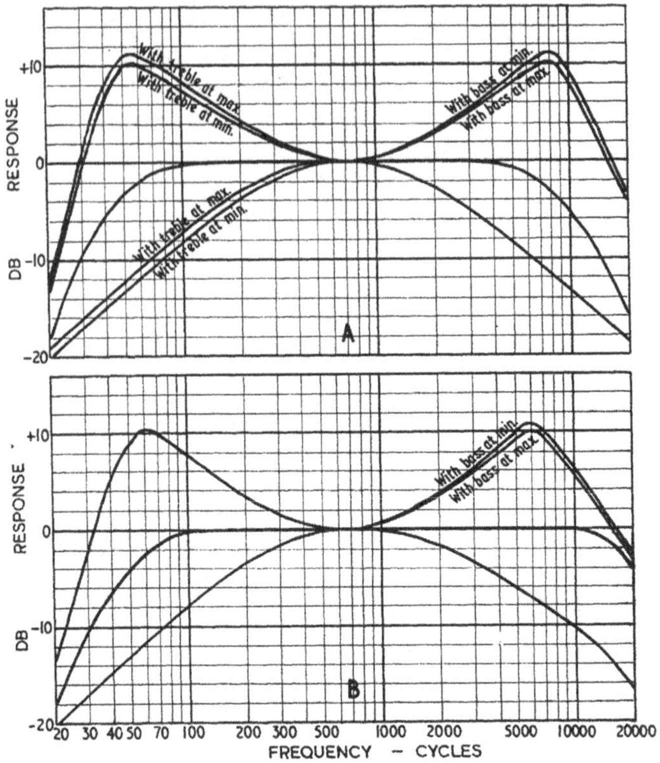

Fig. 1029. *The types of response variation produced by the circuit arrangement of Fig. 1028. The double curves indicate the kind of interaction between controls that can occur.*

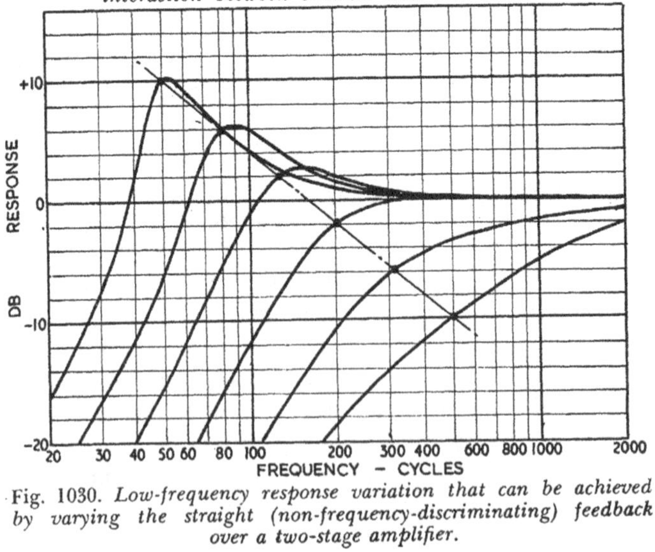

Fig. 1030. *Low-frequency response variation that can be achieved by varying the straight (non-frequency-discriminating) feedback over a two-stage amplifier.*

back purposes is the overall feedback-loop gain. This is achieved by using an internal loop that varies the forward amplification.

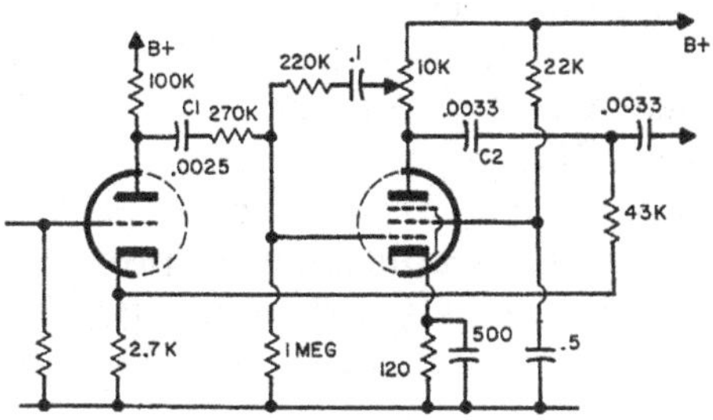

Fig. 1031. *A circuit, using the two parts of a 6AW8 tube to achieve the variation in response of Fig. 1030.*

This inner-loop feedback is taken from the slider of the 10,000-ohm potentiometer in the plate of the pentode section, back into the grid circuit. When the slider is at the top end of the potentiometer, there is no feedback to this short loop. When it is at the bottom end, there is maximum feedback, reducing the overall gain of the amplifier before the overall loop feedback is applied.

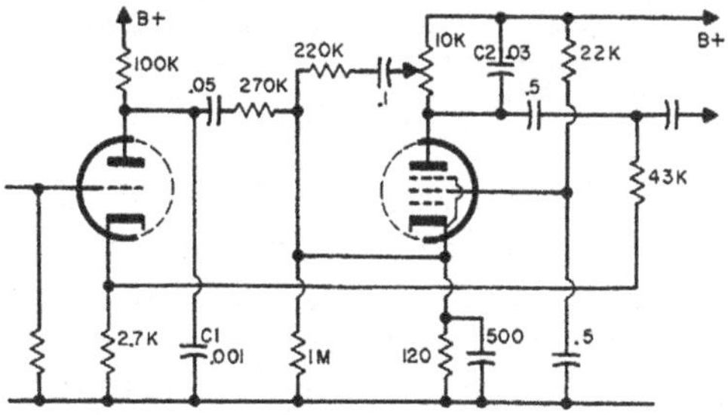

Fig. 1032. *Applying the circuit of Fig. 1031 for producing the same kind of variation in high frequency response. In each of these schematics, C1 and C2 are the controlling reactances.*

The overall loop feedback will maintain the mid-band response practically constant, being essentially dependent upon the feedback factor provided by the resistors. However, the phase-shift effect will radically change the feedback toward the rolloff point, reducing the amount of feedback effective and producing a peak. This will be varied according to the amount of overall feedback, which, in turn, is adjusted by changing the gain of the amplifier by the internal-loop feedback control.

Distortion is minimized under all conditions with this circuit because all adjustment of gain is achieved by feedback, either in the internal or external loops.

The same circuit can be applied to a high-frequency control (Fig. 1032). The difficulty is that the two cannot readily be combined because the response at each end is dependent upon the distribution of the feedback between the inner and outer loops—purely a resistance adjustment. It cannot have two different settings at the same time, one for the high-frequency response and one for the low. The only feasible way to provide independent control of this kind at both ends of the frequency range would be to use the two circuits in cascade, requiring two tubes, one after the other.

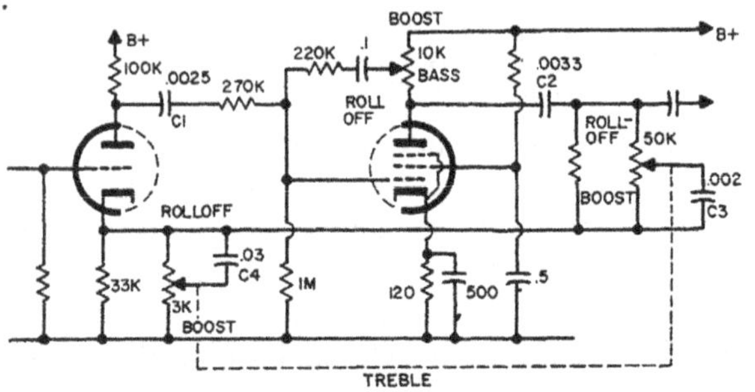

Fig. 1033. *A combination that enables low and high frequency tone control to be combined in the same two stages, giving response variation as at Figs. 1030 and 1001-C respectively.*

An alternative which produces the same low-frequency variation but a high-frequency variation of the more normal pattern, is shown in Fig. 1033. An additional ganged control is provided across the feedback resistors to achieve variation of the high frequency according to the response shown in Fig. 1001-C.

The thing to watch here, however, as in several circuits of this type, is that the high-frequency control does not interact to any appreciable degree with the low-frequency one, or vice versa. It is not usually possible to avoid interaction altogether with this kind of arrangement, but if the interaction can be minimized so that the degree of low-frequency boost, for example, does not change by more than 1 db with adjustment of the high-frequency control, this should be considered satisfactory. This kind of interaction will always occur with feedback type controls more than with the direct type.

Variable-slope rolloff

The final tone control to consider is the variable rolloff, applied particularly for getting the best out of old recordings or indifferently received program material. Two typical circuits are illustrated in Figs. 1034 and 1035.

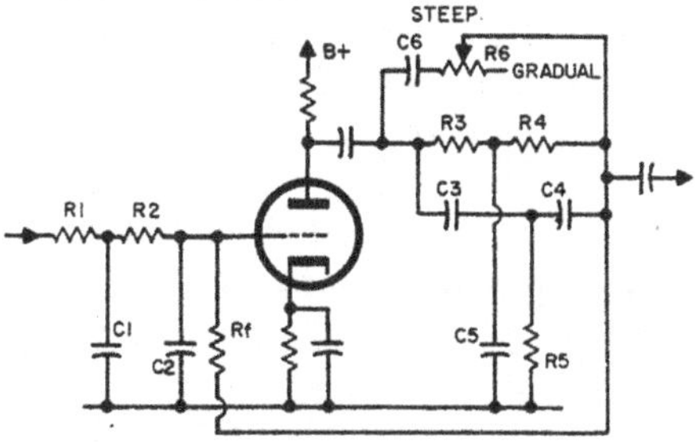

Fig. 1034. *A variable slope rolloff circuit, using a modified twin-T and feedback to achieve the effect.*

The first uses a direct two-stage rolloff arrangement, coupled with a feedback circuit using a modified twin-T network. A twin-T network gives a complete 360° phase rotation. The classic twin-T uses values which produce a null point in the middle of

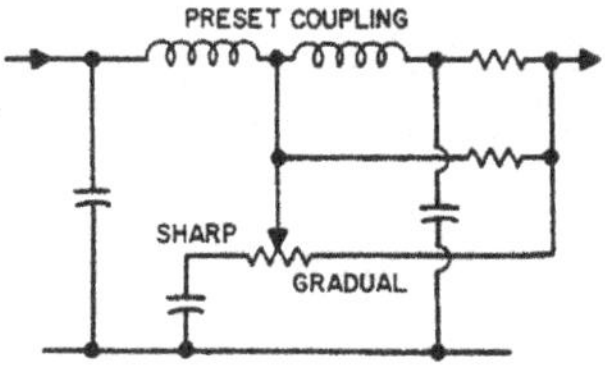

Fig. 1035. *An alternative variable slope rolloff circuit, using inductance, mutual inductance and capacitances to produce the effect in a twin π circuit.*

this rotation. This circuit is employed for straightforward null arrangement to eradicate an unwanted frequency. It can also be employed with a feedback arrangement to produce a sharply peaked response, by arranging 100% feedback over a high-gain amplifier at all frequencies except where the null occurs.

But this is a different application. The twin-T is not the "ideal" arrangement; it is deliberately "misadjusted" to produce a 360° phase rotation. We start out in the mid-range frequency with negative feedback. Then, as the two-stage rolloff starts to produce attenuation, the phase swings round through 180° to produce positive feedback which offsets this attenuation, maintaining a level response out beyond the point where the R-C elements are producing rolloff. Finally, the phase rotation continues through another 180° to produce a very rapid rolloff by reverting to negative feedback.

This is without the control components added to produce the variable slope. Variation is achieved by shunting the top limbs of the twin-T with a variable resistance and a suitable value of capacitor. This changes the rate of phase variation and also the response of the network. By careful choice of components, the overall result is that the same rolloff frequency is used but the slope is changed, producing a response family as shown in Fig. 1005.

Using different capacitor values switched into the circuit as in Fig. 1036, the variable response slope can be achieved at any of a selected number of turnover frequencies. This arrangement forms the basis of the Leak vari-slope arrångement.

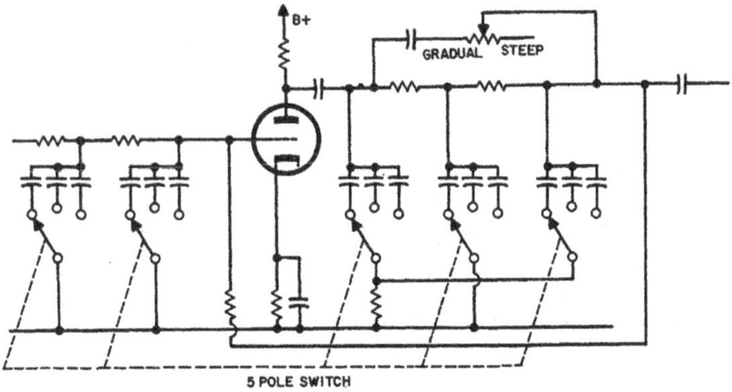

Fig. 1036. *The circuit of Fig. 1034 altered to provide variable slope roll-off at three different frequencies.*

The alternative method (Fig. 1035) employs a mutual-inductance-coupled filter network. The principle is the same except that, instead of using a twin-T with only resistors, capacitors and feedback to achieve the results, this circuit uses a two-stage pi filter network with additional mutual coupling to sharpen the "knee." A resistor damps the center capacitor limb of the twin pi and the other end of the same potentiometer provides additional coupling between this center point and the output of the network. The combined effect of a suitable value potentiometer that changes both these things at once is to produce a variable slope response almost identical with that given by the feedback arrangement.

An important feature of this is the degree of coupling between the two coils of the twin pi network, which has the effect of producing maximum sharpness, accentuating the rolloff point. Also, as the control is changed, it maintains a coupling in progressively reverse action to help continue the response without such a sudden drop toward the high end. This it does by providing some through coupling from the input point to the output point of the filter coils. This circuit forms the basis of the Quad vari-slope filter. Different frequencies of rolloff are again provided by changing the capacitor values in the circuit.

The question now comes as to which of these two circuits will be considered the better for the particular purpose. This is almost impossible to say because the factors which we use as a basis for discriminating between other circuits—that is, the possible distortion and dynamic range—do not apply here. The very reason for which we apply this kind of circuit says we have a poor dynamic range and conditions in which minimum distortion is unnecessary in the program handled.

The combination of arrangements chosen will depend upon the application for which they are to be used. Conflicting factors in selecting an assembly are the versatility and simplicity of any control system. To make the system as versatile as possible means that the response must be variable in a number of different ways. This requires a greater number of controls. On the other hand, this kind of control can more readily result in improper adjustment and thus be misused. A simple control with one knob for the high frequencies and one knob for the low has much to recommend it, especially for those who do not know much more about frequency response than the difference between bass and treble.

So one has to make the choice of a tone-control arrangement according to the use it is expected to get: whether it will be operated by skilled personnel or by nontechnical people who just want to have good listening; also whether it will be used on comparatively high-quality program material which only requires delicate adjustment of balance, or to make the best of inferior program material by getting rid of undesirable sounds.

Transistors

The foregoing variety of circuits has developed with the use of tubes as amplifiers. Now we start all over to apply similar techniques to the more recent innovation, the transistor. This promises to yield an even greater variety of circuitry. At the present "state of the art," the principal application of transistors in audio has been to the miniaturized portable radio and record player field. These scarcely boast of a tone control in the sense used in this chapter: a simple high frequency rolloff suffices.

The important thing in transistor circuits is to remember that *current* is the important feature. For this reason, the usual volume control or tone control arrangement used for tube circuits is reversed. In tube circuits, the input goes to the "top end" of a potentiometer, and the output is taken from the slider. This divides the voltage provided by the previous stage.

In a transistor circuit, the input goes to the slider of the potentiometer, and the output comes from the "top end." This insures that the resistance in the base circuit of the following stage does not change with setting and upset the bias. But it also divides the current provided by the previous stage, rather than the voltage. When the control is turned down, most of the current from the preceding stage (audio component) is shunted through the "bottom end" of the potentiometer, while only a small proportion is fed through the upper part to the following emitter.

This is the volume control arrangement for a transistor amplifier. Tone control uses a variation, with suitably chosen capacitors in the circuit, so that current division, either in the forward amplification or as a feedback function, only takes place in the part of the spectrum to be controlled.

Unlike the tube circuit, in which circuit resistances are principally of importance in avoiding distortion, it is more important to maintain the circuit operating conditions in a transistor amplifier. The dc resistance of both emitter and collector circuits should not be changed by the control settings. The other difference is that the circuits are designed to modify the current division of audio at different frequencies, rather than the voltage division, which is the usual parameter in designing a tube circuit.

Transistors promise to prove much more versatile than tubes. At present only the grounded emitter circuit has been extensively applied. But other connections, and combinations of connections, hold a possibility of greater variety. So it would be impossible to present a comprehensive treatment of this application, compatible with the rest of this book, from the information now available. When it becomes available, it will warrant another book in itself.

chapter

11

overall view of hi fi

THIS book has already run through most of its pages but many things still remain to be said to clarify some of the obscurities that have arisen in audio and high fidelity. The fundamental approach to high fidelity is through an understanding of audio circuits, but you cannot stop at circuits because there is much more to audio than that. However, the biggest region of confusion exists in the matter of circuit choice and the components most closely associated with circuits, which we have covered.

It has been impossible to cover, in this one book, the range of *additional* factors such as, for example, the various mechanical features. A whole book could be devoted—and probably will be—to them: the drive for phonograph and tape systems, the properties of tone-arm and tape-transport mechanisms; different constructions and their relative merits; various approaches to correct pickup tracking on the phonograph; more data about the relative advantages of different transducer types for microphones and pickups, as well as speakers. Then, different ways of constructing tape heads, various circuits for bias oscillators in the case of tape recorders and the question of mixing circuits for recording and choice of the best circuits for tuner design. But in this book we have attempted to cover the things most basic to audio. In conclusion, however, we should include a few words about tidying up what we have; putting it all together, so to speak, and getting rid of the bits and pieces that might be lying around the living room to annoy "the other half."

Much of this part of the story is esthetic rather than a matter of audio performance. At the same time, however, it is necessary to give some consideration to the performance aspect. What we have said in the preceding chapters should help considerably in making many of these decisions.

For example, the chapter on speaker distribution will help you determine how you want to arrange them to suit your home. Deciding this will help you make the necessary choice between a music corner or a music wall to house your equipment (or maybe you will prefer to use a combination, putting the speaker in a corner and having matching cabinetry for the control part that can be placed somewhere along the wall). Whichever you decide on, from the viewpoint of acoustic performance, you can find a suitable cabinet to house your equipment properly. Alternatively, if you prefer to buy a complete package system which fulfills these requirements, we have shown what to look for.

A number of cabinetmakers are in business just to provide cabinets for components. In this way you can buy components made by specialists in each section of the art and fit them into a cabinet of your own choice that will fit in with the decor of your home. In deciding how to mount the equipment in the cabinetry, do not overlook some details about possible performance, from the electrical aspect as well as the acoustic.

Avoid possible hum transfer, for example, between various sources of hum such as the power transformer of the main amplifier or the preamplifier and the pickup and its wiring. Usually the motor board of a turntable is laid out so that the normal position for mounting the pickup will not pick up too much hum from the motor. This is a feature that needs attention and usually has received it from the motor-board designer. But this care is lost if you mount the power transformer of a tuner, preamplifier or power amplifier immediately underneath where the pickup is going to be. A little careful thought here will avoid some tedious hum elimination later.

Then consider possibilities of acoustic feedback from the speakers. It's very nice to get these wonderful low frequencies but they are rather apt to shake the walls and floors of houses and this can result in transmission of vibration back to the pickup or the tubes in the tuner and result in an acoustic feedback effect. Don't have any part of the system that works at low levels, like the pickup, the front end of the preamplifier or a tuner, too close to high-level sound sources so that direct acoustic feedback can take place.

At the same time watch out for mechanical feedback through floors, walls and cabinet housing. Select a method of mounting that provides good absorption of this feedback. Some turntable mountings in custom-built cabinets provide for the motor board to be suspended on foam rubber or equivalent absorptive material. This is a good precaution against acoustic feedback.

Even though acoustic feedback may not be evident without such precautions, they can result in much cleaner reproduction. This is because the first effect due to acoustic feedback is not always a definite howl or vibration. It may result in distortion due to interference with the proper functioning of the low-level circuit or of the pickup. For example, the vibration may tend to lift the stylus out of the groove a little.

Another thing to watch is the position of controls and the lengths of leads that may be involved. Connections between a preamplifier and power amplifier can be as long as you wish, if the preamplifier output has a cathode follower to make it low impedance. But some preamplifiers provide for remote control of the volume and tone control facilities. If this is done by bringing out high-impedance leads to a small remote panel, excessive lead lengths can result in the loss of high frequencies and possibly pickup of hum. It is better, if not convenient to place the preamplifier in a position where the controls are readily accessible, to use a mechanical method of making them remote, such as a shaft extension.

In conclusion we hope that we have not left any readers behind anywhere in this book. I do not use "we" here in an editorial sense. While it has been my objective, as author, to make the material in this book comprehensible to the widest possible audience, the publishers have given every cooperation possible to further this objective. I would like to take this opportunity of extending acknowledgment of this effort in helping to minimize any obscurities I may have left.

Audio has become an extremely complex art (or should it be science?!). It started by being just the sound section of a radio transmitter or a receiver. But modern high fidelity imposes much more rigorous requirements, as this book will have made evident. In sensory units one does not immediately get the impression of something very critical or difficult of achievement. For example, a frequency range (or should it be pitch range?) of 10 octaves and a loudness range of 120 decibels (I suppose 12 bels would be nearer what one would think of in sense units) does not seem so fantastic but, when you stop to realize that, in the acoustic realm, 20 cycles to 20,000 corresponds to wavelengths from 50 feet to 1/20th of a foot, and the intensity range represents a power ratio of 1 followed by 12 zeroes, it seems as though we do have a problem.

On top of this tremendous frequency and dynamic range, our binaural hearing capacity is capable of discriminating, not only between the frequency and intensity of different tones, but also the way in which different waves are radiated. We begin to realize that we have gotten ourselves into a very complex thing.

index

F

G

H

I

L

Made in the USA
Monee, IL
04 November 2026

67384056R00125